DroneAI for Resilient Nature

Mika-Petri Laakkonen
Editor

DroneAI for Resilient Nature

Springer

Editor
Mika-Petri Laakkonen
Oulu University of Applied Sciences
Oulu, Finland

ISBN 978-3-032-19007-9 ISBN 978-3-032-19008-6 (eBook)
https://doi.org/10.1007/978-3-032-19008-6

This Springer imprint is published by the registered company Springer Nature Switzerland AG
The registered company address is: Gewerbestrasse 11, 6330 Cham, Switzerland

Preface

The volume *DroneAI for Resilient Nature* emerges at a pivotal moment in the evolution of autonomous aerial systems and their integration into societal infrastructures. As climate instability and environmental degradation intensify across Europe and beyond, the need for resilient, intelligent, and adaptive technologies has never been more urgent. This publication responds to that call by presenting a curated selection of peer-reviewed research that explores the transformative potential of drones and UAV technologies in environmental monitoring, disaster recovery, infrastructure design, and public service delivery.

The genesis of this volume lies in the collaborative efforts of researchers, technologists, and policy experts who convened around the FinDrones 2024 symposium, hosted in Oulu, Finland. From this gathering, 13 articles were selected through rigorous international peer review, representing the work of over 34 reviewers and contributors from across Europe. The editorial process, led by Principal Research Scientist Mika-Petri Laakkonen of Oulu University of Applied Sciences and University Instructor Johanna Liljeroos-Cork in Tampere University, reflects a commitment to scientific integrity, interdisciplinary dialogue, and societal relevance.

The articles in this volume address a wide spectrum of themes. They examine how autonomous aerial systems are reshaping environmental governance, enabling precision agriculture, and enhancing logistical operations in both urban and rural contexts. Several contributions delve into swarm intelligence and the autonomous coordination of drone fleets, offering insights into scalable solutions for wildfire detection, pest control, and maritime surveillance. Others interrogate the fragmented landscape of European drone regulation, advocating for harmonized and anticipatory policy frameworks that can keep pace with technological innovation.

This publication also foregrounds the societal implications of drone technologies. It considers how these systems can democratize access to services, support sustainable development, and foster institutional renewal. The research presented here challenges conventional paradigms and invites readers to envision new modes of governance, collaboration, and innovation.

We extend our gratitude to the authors, reviewers, and institutional partners who made this volume possible. Their contributions reflect not only technical excellence but also a shared commitment to shaping a more resilient and equitable future. As autonomous aerial systems continue to evolve, we hope this volume will serve as both a reference and a catalyst for further inquiry, experimentation, and impact.

October 2025

Mika-Petri Laakkonen
Editor-in-Chief, Lead/Principal Research Scientist, Oulu University of Applied Sciences, Oulu, Finland

Full Professor, University of Haifa (Affiliated), Haifa, Israel

Acknowledgements

Each article in this volume has undergone a rigorous peer review process, with feedback provided by experts across academia and industry. Thirty-four high-level professional reviewers have contributed to the refinement and clarity of the research presented. Their insights have helped shape the final versions of these papers, ensuring that they meet the high standards of scholarly excellence.

The collaborative nature of this volume is evident in the diverse affiliations and interdisciplinary approaches of the authors. From engineering and computer science to policy and environmental studies, the contributions reflect a rich tapestry of expertise and perspectives.

The editors and contributors of this proceedings volume gratefully acknowledge the support and collaboration of numerous individuals and institutions that made this work possible. Special appreciation is due to the European Commission and its affiliated programs, including Horizon Europe Chips Joint Undertaking, for their strategic guidance. Finally, we thank the authors for their dedication and scholarly rigor and the organizing committee for their meticulous coordination throughout the editorial process.

Contents

Regulatory Frameworks and Policy Analysis

Drone Logistics and Infrastructure Integration

Autonomous Aerial Systems in Societal Transformation: Innovation, Governance, and Integration

Johanna Liljeroos-Cork[1] (✉) and Mika-Petri Laakkonen[2]

[1] Tampere University, Tampere, Finland
johanna.liljeroos-cork@tuni.fi
[2] Oulu University of Applied Sciences, Oulu, Finland
mika-petri.laakkonen@oamk.fi

Abstract. Autonomous aerial systems are rapidly transforming how societies monitor environments, deliver services, and manage infrastructures. This introductory chapter situates the DroneAI volume within these accelerating technological and institutional shifts. As drones and UAVs expand into areas such as precision agriculture, environmental monitoring, logistics, and emergency response, they expose governance challenges that outpace traditional regulatory models and demand new forms of coordination, ethical oversight, and policy capacity. The chapter highlights how autonomous aerial systems act not only as technical tools but also as catalysts for broader societal change, influencing sustainability goals, rural connectivity, crisis management, and data-driven service provision. It outlines the volume's four thematic sections—sustainable innovation and societal impact, swarm intelligence and autonomous coordination, regulatory frameworks and policy analysis, and drone logistics and infrastructure integration—showcasing research that spans engineering advances to policy analysis. Together, these contributions illustrate the multidimensional impact of autonomous aerial systems and their growing role in shaping future infrastructures, institutions, and modes of public service delivery.

Keywords: Drones · Artificial intelligence · Swarm intelligence · Precision agriculture · Airspace governance · Socio-technical systems

The integration of autonomous aerial systems into modern society is accelerating at an unprecedented pace. From environmental monitoring and precision agriculture to logistics and emergency response, drones, and unmanned aerial vehicles (UAVs) are transforming how we interact with and manage our surroundings.

As the integration of autonomous aerial systems accelerates, it intersects with several pressing societal challenges—ranging from climate resilience and rural connectivity to ethical governance and regulatory adaptation. The articles in this volume do more than showcase technological progress; they collectively illuminate how drone technologies are reshaping, e.g., public service delivery, environmental stewardship, and infrastructure planning. From swarm robotics in wildfire response to precision agriculture and rural

M. Laakkonen (ed.), *DroneAI for Resilient Nature*,
https://doi.org/10.1007/978-3-032-19008-6_1

logistics, these innovations also demand new forms of institutional agility, stakeholder coordination, and ethical oversight.

There is a growing emphasis on emerging technologies like AI and autonomous systems posing governance challenges that outpace traditional regulatory models. These technologies operate across sectors and jurisdictions, creating complex socio-technical transitions that require flexible and anticipatory governance approaches. In complex systems, the fragmentation of regulatory authority means that policy innovation frequently emerges at subnational levels, which can foster experimentation but also complicate efforts toward cohesive national or international frameworks. These developments underscore the need for public institutions to build policy capacity, engage diverse stakeholders, and uphold ethical standards as they integrate autonomous technologies into governance and service delivery.

This proceedings volume presents a curated selection of cutting-edge research articles that reflect the diversity, depth, and dynamism of this rapidly evolving field. Taken as a whole, the contributions reflect not only the diversity and depth of technical advances, but also their wider significance for how societies organize, adapt, and innovate in the face of technological change. By moving from methodological breakthroughs to swarm intelligence, regulatory frameworks, agricultural applications, conceptual designs, and questions of sustainability, the volume offers a multidimensional view of autonomous aerial systems—both as engineering solutions and as catalysts for new ways of managing environments, services, and infrastructures.

Taken together, these developments signal a broader societal transformation—where autonomous aerial systems are not only technical tools but also agents of institutional change, policy innovation, and new modes of service delivery. This volume captures that transformation by bringing together research that spans engineering breakthroughs, governance frameworks, and applied integration across sectors and scales. To guide the reader through this multifaceted landscape, the volume is organized into four thematic sections: (1) Sustainable Innovation and Societal Impact, (2) Swarm Intelligence and Autonomous Coordination, (3) Regulatory Frameworks and Policy Analysis, and (4) Drone Logistics and Infrastructure Integration.

1 Sustainable Innovation and Societal Impact

The societal implications of autonomous aerial systems are profound, touching on issues of sustainability, accessibility, and resilience. Chapter "Autonomous Aerial Systems for Sustainable Development of Urban, Rural, and Remote Regions" by Joel Stenroos and Harry Edelman examine how drones can support equitable development across diverse geographic areas. Their work emphasizes the role of UAVs in bridging service gaps and enhancing connectivity in underserved regions. Such examples show how technical innovation can align with broader societal aspirations, reinforcing inclusivity and sustainable development.

Agriculture is one of the most promising domains for drone deployment, offering opportunities for precision farming, resource optimization, and environmental sustainability. In Chapter "Drone Fertilization Experiments and Spraying Outlook in Agriculture", Jere Kaivosoja and collaborators present empirical data on drone-based fertilization techniques. Their work highlights the potential for reducing chemical usage and improving crop yields through targeted interventions. Such applications demonstrate how technological innovation can directly support sustainability goals while also transforming how agricultural practices are organized and managed at scale.

At the heart of this volume lies a strong emphasis on technological innovation. Chapter "Enhancing Drone Operations During Wind Weather: Inertial Measurement Unit and Localized Weather Data" by Rajeeka Hannes and colleagues explores the critical role of localized meteorological data in improving drone performance and safety. The authors delve into the challenges of integrating real-time weather data into UAV operations and propose a robust methodology for enhancing operational reliability. This work is particularly relevant for regions with volatile weather patterns, where drone missions are often compromised by unexpected atmospheric changes. Beyond technical gains, such advances also illustrate how societies can better manage environmental risks and safeguard continuity in essential services through smarter integration of data and technology.

2 Swarm Intelligence and Autonomous Coordination

Swarm algorithms are a cornerstone of autonomous drone research, enabling coordinated behavior across multiple units. Chapter "Persistent Monitoring of Large Dispersed Areas Using Drones Swarm Algorithms and Experimental Validation" by Muhamad Rausyan Fikri and collaborators presents a compelling case for using swarm intelligence in large-scale monitoring tasks. Their experimental validation demonstrates the feasibility of deploying drone swarms for persistent surveillance, with applications ranging from wildlife tracking to border security. Here, technological design also intersects with how societies manage shared resources and safeguard critical areas at scale, showing how coordination among machines can augment coordination among institutions.

Similarly, Chapter "BLE for the Cooperative Awareness of Miniature UAVs: An Experimental Analysis of the Different Advertising Modes" by Sandaruwan Jayaweera and team investigates Bluetooth Low Energy (BLE) advertising modes to facilitate cooperative awareness among miniature UAVs. Their experimental analysis provides valuable insights into communication protocols that can support swarm behavior and collision avoidance in constrained environments. This not only advances engineering solutions but also hints at new forms of coordination and collective action—concepts as relevant for organizations and service systems as they are for autonomous machines.

Furthering this theme, Chapter "Self-organizing Drone Swarms for Detection and Monitoring of Wildfires" by Sebastian Coffeng and colleagues showcases how autonomous coordination can be leveraged for disaster management. The authors highlight the importance of decentralized control and adaptive behavior in responding to dynamic wildfire scenarios, offering a scalable solution for real-time environmental monitoring. Such innovations exemplify how technology not only improves response capacity but also reframes how resilience and collective action are organized in crisis management.

Chapter "Autonomous Drone Swarm Concept of Operations to Support Wildfire Management" by Hannu Karvonen and collaborators further illustrates the potential of drones in crisis response. By developing a Concept of Operations (ConOps) for wildfire scenarios, the authors provide a blueprint for integrating drone swarms into emergency management protocols. This not only demonstrates the transformative promise of autonomous aerial systems for safety and resilience but also highlights the need for new operational frameworks that can connect technology, institutions, and communities in practice.

3 Regulatory Frameworks and Policy Analysis

As drone technologies advance, regulatory frameworks must evolve to ensure safe and equitable integration into airspace. Chapter "Fragmented Implementation of Rules and Regulations for Unmanned Aircraft in Europe?" by Anssi Rauhala and Pekka Leviäkangas provide a critical analysis of the inconsistencies in UAV regulations across European countries. Their findings underscore the need for harmonized policies that can support cross-border operations and foster innovation without compromising safety. This highlights that technological progress and governance adaptation must proceed in tandem, creating opportunities for experimentation but also requiring coordination to sustain trust and effective service delivery.

Complementing this perspective, Chapter "The Future of Autonomous Unmanned Aviation: Bridging the Gap between Technology and Regulation" by Jesse Tolvanen and co-authors explores the interplay between technological capabilities and regulatory constraints. The authors advocate for proactive policy development that anticipates future advancements and facilitates responsible deployment of autonomous aerial systems. Such forward-looking approaches exemplify how technology not only pushes regulatory boundaries but also invites new ways of envisioning how societies manage emerging infrastructures.

Chapter "Standardizing Drone Remote Sensing Data for Agricultural Data Space Ecosystems" by Petri Linna and team addresses the need for interoperability in agricultural drone systems. By proposing a standardized data framework, the authors aim to facilitate seamless integration of drone-generated data into broader agricultural ecosystems, enhancing decision-making and operational efficiency. This emphasis on standards and interoperability reflects not only a technical requirement but also a broader need for cooperation among diverse actors in the agricultural sector.

4 Drone Logistics and Infrastructure Integration

Several articles in this volume push the boundaries of conceptual design and experimental validation. Chapter "Conceptual Development of an Autonomous All-Electric Wing-in-Ground Effect Cargo Craft" by Jussi Aaltonen and colleagues introduces a novel vehicle concept that combines the efficiency of ground effect with the autonomy of electric propulsion. This innovative design has implications for sustainable logistics and low-altitude cargo transport. It illustrates how technological prototypes can also become building blocks for reimagining future infrastructures and the systems that manage them.

In Chapter "Innovative Air Mobility—Key Considerations for Rural Logistics Operations", Vadim Kramar and co-authors explore the unique challenges of deploying air mobility solutions in rural contexts. Their analysis considers infrastructure limitations, community engagement, and economic viability, offering a holistic framework for rural drone logistics. The perspective underscores how technological adoption is always embedded in social and institutional contexts, where community needs and organizational capacities shape feasibility and success.

5 Conclusion

This proceedings volume offers a comprehensive overview of the current landscape in autonomous aerial systems research. The articles presented here not only advance technical knowledge but also engage with broader societal and regulatory questions. Collectively, they illustrate how innovations in drones and UAVs are beginning to reshape how societies monitor environments, deliver services, and organize responses to emerging challenges. From agriculture to disaster management, and from logistics to airspace governance, these contributions show that technological solutions are increasingly intertwined with questions of coordination, inclusivity, and sustainability.

While the emphasis of this volume remains on advancing engineering solutions, the findings also invite reflection on how institutions, communities, and policy frameworks adapt alongside these technologies. In this way, the research presented here demonstrates not only the technical promise of autonomous aerial systems but also their role as catalysts for rethinking how we manage, regulate, and ultimately live with new forms of autonomy in practice.

As drones become increasingly embedded in our daily lives, the insights and innovations captured in this volume will serve as a valuable resource for researchers, practitioners, and policymakers alike.

Sustainable Innovation and Societal Impact

Autonomous Systems for Sustainable Development of Urban, Rural, and Remote Regions

Joel Stenroos(✉) and Harry Edelman

Turku University of Applied Sciences, Turku, Finland
{joel.a.stenroos,harry.edelman}@turkuamk.fi

Abstract. Remote areas, such as the Arctic, face transportation challenges due to long distances, difficult terrain, and hard-to-reach locations. Traditional ground transport networks are costly to build and maintain, particularly for urban, peri-urban, rural, and regional connections, especially in isolated areas. Autonomous and renewable-powered aerial services offer a potential solution to improve sustainability, resiliency, and connectivity in these areas. This study proposes prescriptive advice for integrating autonomous aerial services into remote areas, such as the Arctic, as part of broader urban and regional development efforts. Two exploratory case studies—Bodø, Norway, and Turku, Finland—are examined to further validate prior findings on prescriptive advice on future-proof city development from the viewpoint of emerging aerial services.. The research draws from interviews with development actors and lessons learned from the AEROPOLIS project, examining the Berlin TXL area development project as a preliminary case study. It examines how the results apply to isolated areas with urban centers undergoing intensive urban development activity and includes future visions for sustainable mobility and city development. As the main scientific outcome, the paper outlines factors and approaches impacting the readiness of cities to adopt future mobilities for developing and scaling autonomous aerial services in remote and constrained areas, such as the Arctic cities and regions. The work establishes an approach to investigate local and regional aerial services as a component of holistic aerial transportation modalities. Further, strategies and pathways to adopting autonomous aerial services at urban and regional levels will be discussed. Based on the findings, aerial services can now be better designed as a part of the area development projects to benefit from emerging technologies while securing sustainable infrastructure and investments.

Keywords: Autonomous · Aerial · Services · City · Mobility hubs

1 Introduction

1.1 Drivers of Autonomous Physical Services in the Built Environment

Large-scale city development and urban renewal projects pose settings for implementing and increasing the sustainability and development strategies of cities. Such investments convey opportunities for the change and transition of the cities toward more livable

M. Laakkonen (ed.), *DroneAI for Resilient Nature*,
https://doi.org/10.1007/978-3-032-19008-6_2

and resilient forms of habitats. One field of emerging technologies with the potential to impact cities is autonomous services utilizing swarms of robots. To address this impact, researchers call for more process-based studies on autonomous services and how they may be valuable for practitioners [1]. Furthermore, it is necessary to consider how the strategies of cities and regions can be materialized into solutions and services in city development projects that can benefit from emerging autonomous technologies, such as autonomous aerial services (AAS).

Cities' development strategies are commonly envisioned in the United Nations Sustainability Development Goals (UN SDGs). However, in several cases, such objectives are detached from the concrete localization of the goals, which remain a mere wish list for sustainability [2]. Among the urban development-focused SDGs, SDG 11 includes a task for sustainable cities and communities to make cities and human settlements inclusive, safe, resilient, and sustainable. To that end, addressing transportation with smartness and multimodal transportation solutions, including aerial mobility, has been on the development agenda of several cities globally [3]. This chapter seeks to enhance understanding of the broader ecosystem required to support emerging transportation technologies, as shown in Fig. 1. It explores the key question: Beyond the vehicles themselves, what supporting components are necessary to facilitate the adoption of autonomous aerial mobility? Additionally, it examines insights from innovative urban development projects to offer practical guidance for planning and implementing sustainable, future-proof transportation solutions.

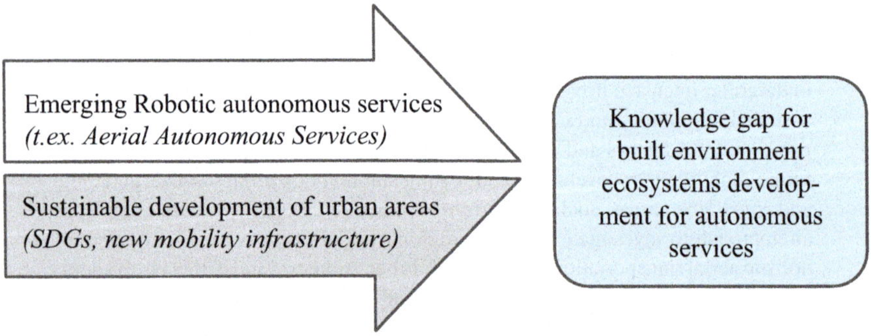

Fig. 1 Knowledge gap in leveraging autonomous services in support of sustainable urban development as the key theme of the research

1.2 Adoption of Sustainable Transport in Urban-Rural Environments

The strive for autonomous services in the built environment, including unmanned aerial vehicles (UAVs), is based on three rationales: (1) The technological push caused by the technological transition [4], (2) the robots provide support services, well-being, and new types of interaction within said services [5], (3) the multimodal solutions enable autonomous service ecosystems with complementary components that support multiple commodities or service processes, for example, as a hybrid service of air and ground transport for last-mile deliveries [6]. The three rationales have been summarized in Fig. 2.

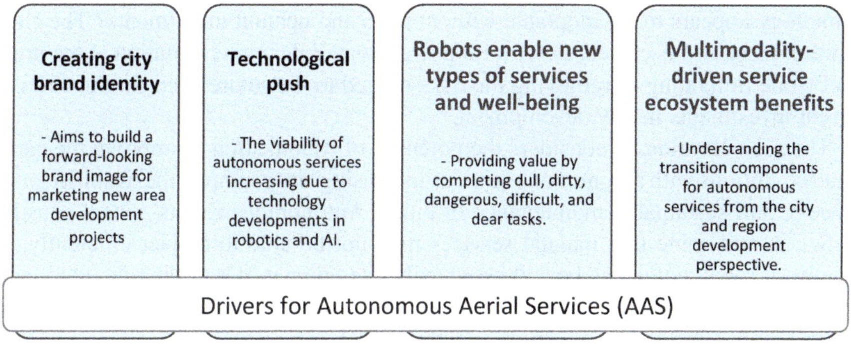

Fig. 2 Drivers for UAV-based autonomous aerial services

The first rationale exposes the cities to technological development. From a theoretical point of view, such development can be investigated and better understood as a process of technological transition (TT) [7]. Within the transition, one technology is replaced by a new evolving and novel technology that gains a foothold in so-called niches, then forms larger clusters of users as regimes, and finally defines the entire landscape of operations, thus changing the paradigm [8]. One example of a paradigm-level technological transition is the introduction of automobiles that began replacing horses and carriages on the street. This transition was an incremental and accelerating process of sociotechnical transition, where cars became the prevailing landscape on the streets [9].

New evolving technologies, such as autonomous systems, provide a potential evolving niche at the technological level that has the potential to transform the paradigm or status quo in which city services and operations are now provided and coordinated. Autonomous Aerial Services offer the potential to transform private and public services into urban environments, for example, by creating new ways to implement point-to-point customer deliveries, decreasing the first response reaction time, and providing situational awareness, such as monitoring the environmental conditions and the safety of cities. The autonomous features supporting the wide adoption of these services are currently limited in practical applications in the built environment, but the feasibility of autonomous features is growing rapidly due to the technological advances in robotics and AI. Therefore, autonomous aerial services are expected to be available within 10–15 years [10]. The technological transition will, in turn, impact the modalities of transportation. To that end, future aerial services are already having an impact on city and regional development plans and objectives, as sustainable design solutions for infrastructure and the built environment must be adaptable to new emergent technologies.

Considering the global competition to which countries and their cities are not immune, the pressure of being at the forefront and the fear of missing out can be seen as one of the transition drivers. Now, the critical question is to what extent the actors behind the drivers have the capacity and will to materialize into tangible efforts, such as designs for future-proof city developments and investments, which may foster the adoption of both aerial and other autonomous services at large over a period. Or are the autonomous systems merely on the wish list for place-making and marketing purposes to create an image of progressive and sustainable development that may happen if future technology appears to be adoptable without risks and upfront investments? The choice between these two is critical to set forth a strategy and its implementation at the project level while mitigating or removing the risks related to the future needs of the cities and current investments in city development.

The second rationale considers the potential of robotization to improve the current urban conditions with autonomous services increasing, for example, the livability, attractiveness, and sustainable maintenance of cities. Autonomous robots, which constitute AAS, can add value that manual services may not be able to deliver efficiently. The benefits of robotization can be addressed using the popular 5Ds of the robotization narrative, as they can provide value by completing tasks that are dull, dirty, dangerous, difficult, and dear [11]. In resource-constrained environments, AAS offers a practical alternative to manual services. Powered by renewable energy sources such as hydrogen fuel cells or electric batteries [12], these systems contribute to cleaner transportation by reducing greenhouse gas emissions and air pollution. Beyond environmental benefits, AAS promises improvements in operational efficiency, safety, and cost-effectiveness, making it an asset for large-scale urban maintenance and service delivery at the large scale of cities.

The third rationale ties the first two rationales together, as there exists a pressing need to understand the components of transition for autonomous services from the city and region development perspective, connected to the overall needs of a city or region. Multiple different components (e.g., vehicles, data, and physical infrastructure) and mechanisms (e.g., regulation and business ecosystems) are needed for autonomous integrative solutions.

There is a significant disparity in how robotic services are regulated, depending on their mode of operation. Aerial robotic services are subject to aviation regulations, which are traditionally strict, emphasizing safety and standardization. In contrast, land-based autonomous or remotely operated robots have enjoyed relatively permissive oversight, often operating freely on urban sidewalks. This regulatory imbalance presents challenges for scaling aerial robotic services but also safety hazards for ground robotics, such as delivery robots occupying open space to a growing extent.

For AAS to become scalable, they must operate autonomously with minimal human intervention. Achieving this vision requires continuous, autonomous control and integrated fleet management across both airspace and ground operations. As our analysis focuses on the European context, from the European perspective, scalability is not only a technical requirement but also a strategic objective aligned with broader EU goals. As an example, the EU Drone Strategy 2.0 emphasizes the need for a smart and sustainable manned and unmanned aviation ecosystem [13]. Europe has already initiated efforts to support this transition, one notable example being the development of U-Space, a traffic management system for unmanned aerial vehicles [14]. Furthermore, the European Union Aviation Safety Agency (EASA) is working to harmonize aerial service regulations across member states to support broader autonomous UAV use [15].

However, as aerial services expand, particularly in densely populated areas, there is an increasing need for localized airspace management. To that end, some cities have begun to take initiative by introducing their own zoning and regulatory frameworks for low-altitude airspace, rather than relying solely on national aviation authorities [16].

For the scope of this chapter, we do not delve into the regulatory details but focus on the technological and urban development aspects of AAS within the European context, considering Central and Northern European regions.

Besides regulation, components like physical infrastructure can be argued to be critical for establishing the needed AAS. However, these components and mechanisms may vary, and for example, the initial aerial services may be offered at a limited scale, based on digital solutions, or operated partially manually. To that end, to ensure the materialization of development efforts, a clear picture of the initial key components enabling the transition to autonomous services is needed.

As a starting point for the common approach and design strategies, some existing development trends can be expanded to serve the needs of the transition to autonomous mobility services. One example of these trends is the implementation of mobility-centered service buildings called *mobility hubs*, which incorporate multiple forms of modalities as transit access points, such as parking, shared economy vehicles, and potentially autonomous vehicles, and interaction points for them [17]. The mobility hubs can be designed to include a variety of use cycles with a minimum transition to future aerial and autonomous services from temporary uses before autonomous service maturity. Furthermore, beyond a single service function or mode of transportation, the components must be able to function as connector points for a variety of mobility needs and applications. In summary, new design approaches are needed for resilient mobility solutions and their scalable impact within the EU. The connector points, such as the mobility hubs, need to be studied to create a framework for urban development for autonomous services that benefit beyond urban life, rural and regional development, and connectivity. To that end, a clear vision for multimodal mobility solutions in large-scale area development projects should be outlined to drive area development and transition projects with a resilient and common urban development strategy. As there is currently a lack of discussion on comparable current urban development examples, this chapter aims to address the need for a better understanding of the available solutions, strategies, and their integration into urban development projects.

2 Components and Drivers of Transition to Autonomous Services

2.1 Autonomous Robotization Extending Urban Service Capabilities

In urban environments, there is already a range of robotic services and examples of autonomous robotization to meet various needs. Ground-based robots are used for delivery, lawn mowing, construction works such as painting, wall assembly, and bricklaying, maintenance such as window cleaning, infrastructure works, construction monitoring, and more. Aerial robots are well-suited for providing last-mile deliveries, data collection, and aerial manipulation to hard-to-reach or dangerous locations, such as windmills [18]. Despite the existence of robotic services, several appear to be lacking. Robots, for example, may be used to manipulate the built environment more extensively, allowing for autonomous maintenance of the built and green environment.

The emergence of robotic services leverages urban environments in new ways on both small and large scales, creating opportunities for both synergic collaboration and conflict. For example, ground-based robot deliveries give customer deliverables by leveraging curbsides and pre-existing road networks. This may be seen in the more efficient use of roadways in terms of utilization, but it also opens the possibility of conflicts with robotic vehicles, other vehicles, and people [19]. On the other hand, aerial robots employ airspace less congested than ground level, providing opportunities to execute services without adding ground congestion [20].

The possibility of employing robotization in an urban context raises concerns about the need for a more comprehensive understanding of how the built environment is used to promote the transition to autonomous robotic services. Currently, development is driven by privately owned companies, and it is typically focused on a single service. In contrast, public development organizations respond to emerging development and act based on current interests [21]. To that aim, a broader systemic perspective that considers autonomous robotization through the lens of networked robots that utilize the built environment efficiently yet collaboratively is required. The inclusion of aerial robotics is also required to establish a robust approach in terms of emerging future mobility services. Furthermore, aerial mobilities can play a key part in expanding the urban service capabilities of networked robots by enabling aerial operation, hence implementing multimodality. The possibility of employing robotization in an urban context raises concerns about the need for a more comprehensive understanding of how the built environment is used to promote the transition to autonomous robotic services.

2.2 Current Development of Vertiports Located in the Broader Urban Context

To understand the role of the built environment and robotization infrastructure as potential drivers of mobility transitions, it is critical to view the current developments in urban air mobility (UAM) and aerial services at large, including diverse scales of transport in urban, peri-urban, rural, and regional settings.

The concept of the *vertiport* has emerged as essential ground infrastructure to support aerial services, particularly for large-scale deployment [22]. While not all such services are autonomous, they require dedicated facilities for take-off, landing, and fleet management. Companies like Uber, in collaboration with Foster + Partners, have proposed hub-like systems for UAM [23], while Skyports Infrastructure is repurposing containers [24], heliports and rooftops in London for future UAM operations [25]. In logistics, Amazon envisions regional hubs like the Hive, which forms a regional transportation hub and connection to heavy ground transport [26], and an airborne fulfillment center [27], while other companies, such as Matternet, Zipline, and Wing have introduced more lightweight delivery models using lockers [28], drop zones [29, 30], and tether systems [31]. These innovations primarily focus on the functionality of aerial platforms, often treating them as standalone systems or swarms. However, there is a notable gap in integrating these services into the broader urban and regional fabric. A more holistic approach, such as embedding aerial services within mobility hubs, can bridge this gap, ensuring that aerial mobility complements other transport modes and contributes to livable, interconnected urban and rural environments.

Despite all the research and development activities focusing on commercial aerial logistics, there has been very limited attention on the holistic approaches to addressing aerial services as a system and part of an urban system and network. Rather, the approaches, such as Amazon's hive, have been debated as nightmares to urban planning due to the ignorance of the overall functionality of logistics and impacts on the overall built environment [32]. As another example, a holistic approach was proposed for Berlin's ICC, where activists envisioned transforming it into a Mobility Innovation Center featuring autonomous aerial and electric ground transport [33].

2.3 City Development Trends for Utilizing Autonomous Services

Autonomously operating services can establish new urban service chains and networks, reshaping how the urban environment is used and perceived. The mobility hub approach can be used to implement a multitude of autonomous services, integrating necessary service and maintenance infrastructure, for example, for vertiports, and impacting the sociotechnical transition on larger scales.

Given that autonomous services, such as AAS, are still developing and emerging, the literature has several approaches to the issue of the interaction between services and the urban environment. The subject of robot-city interaction has emerged to examine a variety of perspectives, such as navigation from point to point, perception of the environment, management of the actions of humans and robots, manipulation of the environment, verbal communication for interacting socially, and acquisition of new competencies by interaction in the city [34]. The concept of an autonomous city has emerged to cover a wider city approach that also considers autonomous services in a broad urban context [35]. The autonomous city approach has the potential to provide resource-efficient services that considerably increase the quality of life while also enabling new forms of lifestyles. Improved shared mobility without drivers, as well as last-mile delivery, allows services to be offered at previously impractical scales and locations. For example, rural locations have lower service levels in mobility and commerce, which autonomous services can compensate for by linking to neighboring urban areas via unmanned autonomous services. Furthermore, autonomous services can enable multilocal lifestyles by enhancing service capacities and enabling novel lifestyles in remote regions such as rural and peri-urban areas. Thus, the autonomous city can also extend its service capabilities to benefit the surrounding region to support multilocal lifestyles. However, the approaches of robot-city interaction and autonomous city have still been lacking consideration for aerial robotics, requiring an extended perspective of the urban development potential.

Emerging mobility trends can be recognized in the development of cities and regions as an increasing interest in alternative mobilities for service-based future urban development. To that end, urban development projects based on autonomous services have been proposed for implementation worldwide. Such examples are Woven City (autonomous systems as part of city development) [36], Seoul (urban air mobility) [37], Dubai (world's first permanent commercial vertiport) [38], Helsinki (Emergency UAV services) [39], Berlin TXL (Multiple mobilities such as UAVs included in strategic planning) [40] and emergency health transport pilots in Belgium [41].

From the perspective of sociotechnical transition, these projects present the formation of initial regimes for utilizing autonomous urban services. For covering the potential to utilize AAS as part of urban development, the Berlin TXL project and its mobility hub approach are covered next, along with observations on ways to facilitate autonomous mobility transition in cities and regions.

3 Materials and Methods

3.1 Methodology and Approach for Case Studies

This paper uses a mixed-methods approach, drawing on case studies of city development projects in Berlin, Germany, as well as case studies from Bodø, Norway, and Turku, Finland. Given the complex nature of the transition to radically innovative mobilities, the case study approach was chosen to provide a comprehensive and general overview of urban development activities and their relationship to the transition to AAS. Further, due to the availability of comparable projects, we have limited the perspective of the study to consider mainly Northern and Central Europe. As a unit of study, the drivers and readiness of the built environment solutions to support the future-proof transition to autonomous aerial services are analyzed for each of the urban development projects. To that end, we aim to answer the following research questions: Aside from the aircraft, what other sociotechnical components of the system of systems, such as ground infrastructure and other urban elements, are required to enable the transition to autonomous aerial services? Furthermore, what can be learnt from ongoing city development initiatives with unique transport designs to provide prescriptive development advice for the transition to sustainable autonomous services?

The paper employs an exploratory, embedded multi-case study approach and qualitative methods. The case studies of Bodø and Turku are analyzed based on the critical analysis of the Berlin TXL case materials, and underpinning the theoretical foundations of the AEROPOLIS project for integrating unmanned aircraft infrastructure in an urban environment [42], and then complementing and critically applying the analysis to the Bodø and Turku cases and externally validating the findings of a preliminary case study on the guidelines for utilizing autonomous service transition.

The introduction of mobility hubs as part of area development, which includes aerial mobility as well, is being explored as a major driver of the transition to autonomous services. The Bodø case study is used in comparison as a similar development to that of Berlin, involving transforming the former airport area into a hybrid area with new mobility functions. In contrast, the Turku development project does not have drivers for a specific aerial mobility but rather is used to critically explore the feasibility of offering autonomous aerial services alongside other developments without conventional aerial infrastructure, such as airport projects.

The primary data sources include semi-structured interviews with urban development stakeholders in Bodø and Turku. Secondary data sources include interviews with experts in Berlin with development documentation of the plans, assisting in comprehending the development goals and how AAS correspond to those aims. Thematically, the sources are analyzed from the perspective of sociotechnical transition, considering possible niches to initial regimes by identifying existing drivers such as development strategies, the ability of built environment solutions such as mobility hubs to adapt to the transition, and the organizations' readiness to enable the transition to AAS as part of broader development projects. Primary sources, such as interviews, are supplemented by secondary sources, such as documentation analysis, to generate a comprehensive identification of the critical objectives for urban growth and AAS transition.

3.2 Berlin TXL Development

The primary case study covers the former Tegel Airport development in Berlin, transforming into a sustainable urban district called Berlin TXL. The project comprises three key areas: the Urban Tech Republic, Schumacher Quartier, and Tegeler Stadtheite Park. Urban Tech Republic industrial park in the former terminal area will employ 20,000 people and host 5000 students [43]. Schumacher Quartier is planned as a climate-resilient residential district on the eastern side of the airport. This area will offer 5000 apartments for 10,000 residents, with a focus on sustainable living. Construction starts in 2027, with completion expected by the mid-2030s [44].

The development emphasizes five climate goals: energy-efficient buildings, sponge city design for rainwater management, local timber construction, biodiversity support, and a car-free environment with a mobility hub approach providing shared transport options. The mobility hubs in the district are strategically placed on its edges, evenly spaced to link efficiently with public transport, bike paths, and pedestrian lanes. This network aims to support a car-free environment, promoting a central area focused on pedestrians and cyclists. The mobility hubs will not only provide parking but will also serve as hybrid spaces with commercial services, shared vehicles like cars and scooters, and potentially, facilities for UAV infrastructure acting as a core infrastructure for potential autonomous aerial services.

As part of the AEROPOLIS project in Berlin, approaches for making mobility hubs future-proof with autonomous mobility and infrastructure were investigated, particularly those involving flying vehicles. Based upon several stakeholder meetings and expert interviews, the study developed speculative design proposals for various forms of infrastructure to address the potential for mobility transition. These proposals include adaptable mobility hubs that may host aerial service infrastructure, such as payload and energy source handling, on the rooftop and uppermost floors, connecting the services to ground floor service interaction points, such as parcel offloading and autonomous people mobility terminals [45]. These approaches offer an urban development-based perspective to enable the transition to autonomous aerial services. To that end, the Berlin TXL development has the potential for providing autonomous airborne services: It already has a history of aviation and aerial services, and the area development has ambitious plans for integrating numerous modes of transportation, enabling autonomous services. Furthermore, the speculative concepts created in Berlin have the potential to facilitate a broader conversation about urban development capacities in response to the autonomous transition of aerial services.

3.3 Bodø Ny by—Ny Flyplass Development

Bodø is transforming airport property into a new city district, called Hernes. Bodø's new airport project aims to improve infrastructure and promote sustainable urban development. The city of Bodø has a strong cultural and historical connection to aviation, having served as a center for aerial authorities and operations. When completed, this region will encourage sustainable urban living and become a zero-emissions zone, housing 30,000 residents and creating 20,000 jobs [46]. Avinor, a state-owned airport operator, plans to extend the airport 1500 m to the sea, freeing up 300 ha of land near the city center for housing, businesses, and modern infrastructure development. The first plots will be available in 2025, with full development anticipated by 2030. Bodø Municipality, with Nordland County Council, oversees planning, aiming for a future-proofed district that responds to the city's need for commercial space and supports economic growth. The project intends to reduce environmental effects through energy-efficient building materials, garbage recycling, and eco-friendly transportation choices, contributing to Bodø's carbon reduction goals. Furthermore, it intends to build a high-capacity facility with an annual passenger capacity of 1.9 million, ensuring that the infrastructure supports future growth and integrates with local and regional public transport, connecting the Bodø area regionally and internationally, and promoting reduced car dependency.

Bodø's new airport project mirrors Berlin's TXL development goals by transforming the previous airport and aviation-based region into a multifunctional area. Furthermore, the isolated geography of Bodø and its surroundings implies potential benefits for using aerial services.

3.4 Turku Kupittaa Development

Kupittaa area in Turku, Finland, is developing under multiple subprojects. The Kupittaan Tiedepuisto (Turku Science Park) development project is a flagship initiative spearheaded by the City of Turku, aimed at fostering innovation, economic growth, and sustainable urban development of the city. The primary goal of the project is to develop a former industrial area into a mixed-use area with 16,000 new residents and 18,000 new jobs to establish Turku Science Park as a leading innovation center that attracts a dynamic mix of businesses, universities, and research institutions [47].

A major component of the project is its commitment to sustainable urban growth. The Science Park emphasizes the integration of green infrastructure, improved public transit access, and energy-efficient buildings. This vision not only focuses on creating an environmentally friendly space but also aims to establish a model for urban areas that prioritize sustainable design and operation. Enhancing the area's connectivity is another essential goal, as the development seeks to improve links between Kupittaa, Turku's city center, and broader regional transit systems, ensuring accessibility for residents, employees, and visitors alike. Additionally, the project envisions a community-oriented environment that fosters inclusiveness, collaboration, and a high quality of life, creating a space where residents, students, and professionals can thrive. The project's commitment to sustainability aims to minimize its carbon footprint through green building practices, serving as an example of environmentally conscious urban development.

In contrast to Berlin and Bodø, the Kupittaa development in Turku does not draw on past aviation operations, but rather strives to promote mobility on many levels, both local and regional. Because Turku is located near the Archipelago Sea, which consists of numerous small island-based communities, it has the potential to improve connections between isolated areas by providing aerial services comparable to Bodø.

4 Results

4.1 Mobility Drivers and Readiness of Urban Infrastructure to Address Emergent Mobilities and Services in the Case Area Developments

In all case studies, the drivers for developing new mobility services are linked with key area development goals, and mobility has been closely associated with new development. Other common core strategy topics for development projects used by the selected case studies involve developing a smart hybrid district, including residential and commercial functions, as well as improving the natural environment. Table 1 presents key findings about the strategic priorities of the area development case studies.

Table 1. Drivers and barriers of the cases to transitioning to autonomous services and development goals

Case	Berlin Berlin TXL, Schumacher Quariter and Urban Tech Republic	Bodø Hernes, New City—New Airport	Turku Science Park, Kupittaa area developments
Development size indicators	• New residents: 10,000 • Jobs: 20,000 • Building area: 2, 480,000 m^2	• New residents: 30,000 • Jobs: 20,000 • Building area: 3,000,000 m^2	• New residents: 16,000 • Jobs: 18,000 • Building area: 900,000 m^2
General Key Strategic Goals	• Utilising mainly timber construction • Effective rainwater management, sponge city approach • Encouraging green mobilities and reducing private car usage • Supporting biodiversity [48]	• Green structures adaptive for climate change • Preserving existing cultural landscape • A smart, attractive and compact city district • Promoting the well-being of the diversity of residents • Environmentally friendly transport network [46]	• Sustainably growing district • Attractive Centre for smart mobility • Cluster for experimental innovation • 24/7/365 living centre for encounters [47]

(continued)

Table 1. (*continued*)

Case	Berlin Berlin TXL, Schumacher Quariter and Urban Tech Republic	Bodø Hernes, New City—New Airport	Turku Science Park, Kupittaa area developments
Selected Mobility hub Approach	• Five evenly spaced mobility hubs with variable focus on different mobilities [49] • Strategic planning including conventional and emergent mobilities [45]	• Ten mobility hubs dispersed around the area with comprehensive hybrid solutions consisting of multiple mobilities [50]	• Reservation for one centrally located mobility hub with conventional mobilities and mostly parking [51] • Mobility infrastructure is decentralized around the area [52]
Preparedness to respond to innovative mobilities	• Future-proofing approach, open to innovative mobilities [45]	• Development as a transition of 50 to 100 years, open to innovative multimodal mobility, need for safe transition [53]	• Innovative mobility is not a part of key strategic development, however, has been discussed internally for visions and proof of concepts [52]

For Bodø, the plan has the potential to integrate emerging solutions for mobility into urban development as the plan has 10 mobility hubs along the development area. Furthermore, the approach of using multiple mobility hubs does not directly include future forms of mobility, but includes multiple applications like microtransit, rideables, scooters, and neighborhood vehicles, and uses the strategy of flexibly preparing for possible unexpected developments in the future [47]. To that end, emerging autonomous vehicles are not currently part of the mobility plans, but according to the interviews, there has been strategic discussion regarding their inclusion.

Despite the possibility of utilizing mobility hubs for aerial mobilities, the transition in Bodø is both conflicted and enhanced by the connection to the airport development. The airport itself sets some barriers for buildings, such as building height limitations [54]. Further, conventional airport operations require airspace, which, in turn, with current regulations, cannot be simultaneously used by unmanned aircraft providing autonomous aerial services. To this end, the infrastructure and aerial missions were conceived to be located a safe distance from the airport while benefiting from the services of the manned airport ecosystem, such as green fuels. In contrast, the airport and the current aerial ecosystem in Bodø carry different synergies with an already existing acceptance culture for flying vehicles and technical, regulatory, and managerial expertise for aerial services to enable the safe transition from aerial services to autonomous emerging solutions [53].

In Turku, the mobility hubs are considered a more general concept, supporting general mobility developments like rail, public transport, and seaport development [55]. This approach lacks a more defined take on what different forms of mobility the mobility hubs may include. Furthermore, the Kupittaa area development includes only one mobility hub that has been indicated to function as a hybrid parking space [51]. According to the interview, the clear driver for development in Turku is the inclusion few critical components, such as the urban tunneling project of a highway with a deck connecting opposite sides of the highway, along with the large-scale educational complex Taito-campus [52]. In addition, the initial location of the mobility hub causes challenges for airspace management because of the planned high-rise buildings nearby.

4.2 Opportunities to Extend Urban Living through Autonomous Services

For both cases in Bodø and Turku, the potential to utilize mobility hubs and built environments exists to adopt the autonomous aerial services to enable the sociotechnical transition. However, the current approaches used in the development vary, thus affecting the possibility of utilizing the prospects for autonomous aerial services.

For Bodø's New city—New Airport project, the large number of mobility hubs and ecosystem benefits creates a potential to deliver flexibly different services that can be used to enable strategic development goals, such as promoting well-being and attractiveness through new aerial services. Bodø acts as a hub for logistics, combining aerial, ground, and sea mobility and logistics [53], and from peri-urban and regional perspectives, this opens new opportunities for nearby areas not covered with services, one instance being the Salten region with a population of 73,000. Aerial services operated from the new airport area could extend deliveries of consumer goods and medical equipment, raising the public service capacity in neighboring rural areas. Furthermore, from a broader regional perspective, the County has 11 traditional airports, posing a question for the future development of aerial services [53]. The possible utilization of autonomous aerial services that are more flexible from the infrastructure point of view can reduce the need for conventional airports such as the ones in the county, if the innovative solutions replace the traditional ones. However, the current regulation separates autonomous UAV operations and conventional manned aviation, causing a potential for conflicts in the case of the implementation of autonomous aerial services at airports. However, the regulatory framework for sharing and using the airspace is underway in the European Union through the efforts on the U-Space management, that is, common rules and protocols for the shared airspace [56]. Therefore, along with the increased and trustworthy automation of airspace monitoring, the co-existence of manned and unmanned aircraft is likely to take place within the development timespan of the large urban development projects, such as Bodø. On a national level, the implementation of AAS can be used to leverage the region's and country's arctic conditions as a testbed for aerial services. The weather conditions, such as wind, frost, and rain, cause challenges for aerial vehicles, and creating services resilient enough for arctic conditions could help create generally more resilient aerial solutions. The region's vast distances to meet transport demands for people and products, along with climate-related disturbances that cause temporary isolation of areas, necessitate solutions in which aerial services could play a crucial role [53].

For Turku, there exists a potential for autonomous aerial services to extend the strategic points of Kupittaa development. The goal of connecting two sides of the city with and possibility to use aerial services to connect areas beyond the conventional road and light rail networks poses potential for future area development. This can enable more extended multimodality that is also driven strategically in Turku [55], by creating travel chains that combine both aerial and ground mobilities. Furthermore, several emerging transitory niche prospects were recognized, as there had been internal discussions about cooperation possibilities for aerial deliveries from a neighboring central hospital. This approach is strongly related to early services that may be linked to first responder actions, providing a possible niche to push the transformation in Turku. However, in terms of Kupittaa development and developing forms of mobility, there is a lack of system-level thinking, which calls for a framework for the design and development to enable the scaling of aerial services in the future. Furthermore, in addition to connecting different districts of the city, the research unveiled the potential to support surrounding local communities in the Archipelago Sea, increasing the impact of implementing AAS [52].

4.3 Guidelines to Enabling the Transition to Autonomous Aerial Services for Urban Development

Next, guidelines for developing and scaling autonomous aerial services in remote and constrained areas are covered, considering the drivers of the transition and creating directions to overcome possible barriers. The guidelines have been created by applying learnings from Berlin to Bodø and Turku, revealing certain similarities.

Strategically, all three development projects have been considered to include radical forms of mobility, such as autonomous aerial mobility at some level for the development, but the stakeholders' awareness and commitment to implement it vary greatly. Considering that the dominant use cases for AAS in urban environments are still emerging, the most flexible way to utilize them can be starting with simple use cases and broadening up through the development of regulations, technology, and practices. Furthermore, all three development projects utilize the mobility hub approach. However, their approach and focus on using these hubs for varying mobility, including aerial services, creates a critical barrier to the adaptability of mobility hubs for other uses, such as enabling the use of autonomous aerial services in the future. In addition, the flexibility aspect also enables multimodality, including other forms of autonomous mobility such as ground rovers, which we argue to be critical for creating impactful aerial autonomous systems that are scalable. The placement of infrastructure, such as mobility hubs, should also consider locations that are close to services and clients so that the offered services are still appealing, supporting the integration of autonomous aerial services. Furthermore, the planning should also consider the regulation for providing aerial infrastructure, such as vertiports, so that they can be implemented with vertiport guidelines, leading to the need for dedicated airspace that is not limited, for example, by neighboring tall buildings. Consequently, opportunities in close urban proximities and the potential for enabling services regionally should be considered as well.

To summarize the findings, the guidelines for facilitating autonomous aerial transition make up three key points:

1. **Leverage the local context, opportunities, and unique features to root transition activities**

In all cases, the one critical feature of the potential for facilitating the transition to AAS is derived from the local context. Applying technological transition theory, this can be viewed as emerging niche fields that aid in the process of transition. For Berlin, the drivers from the local context were the old airport with a history in aviation and a future-proof-oriented vision with goals to utilize aerial mobility to create a car-free district. Bodø's strong aviation culture and airport ecosystem with knowledge and expertise enhance the integration of services into future development. In Turku, the local context emerged from identified internal discussions based on work experiences that capitalized on synergies in the local environment, such as intentions to start cooperation projects for medical drone deliveries with the surrounding hospital district. This localness, as well as the unique features that come with it, can and should be recognized and used to tailor potential autonomous service offerings and development plans.

2. **Provide temporal flexibility in the mobility infrastructure to support future-proof projects**

One of the findings from the Berlin TXL development was that temporal solutions and transitory applications should be examined to enable more inexpensive and appealing design from the developer's perspective. Furthermore, this can support the adaptability for strategic future needs in Bodø and Turku. For example, spatial reservations for emerging autonomous aerial mobilities can be made when designing and programming mobility hubs, and the requirements of those services should be considered during planning to ensure a smooth transition to emerging mobilities in the future. However, these spatial limitations provide an opportunity to establish flexible spaces that can be employed for multiple uses and functions besides AAS before emerging mobilities necessitate the spaces. For example, the manned autonomous aerial mobility hub was designed to function initially as a sports area and, in the future, as a terminal for urban aerial mobility.

3. **Engage diverse potential beneficiaries to ensure all critical strategic factors are addressed for the planned transition**

Urban development project strategies play a crucial part in future-proofing thinking and leveraging emerging mobilities. Stakeholders for supporting development from a futures-oriented viewpoint, such as researchers, experts, and the community, should be used to extend the future-proofing of the strategy for validation and mitigating the risks associated with the plans. For example, Berlin TXL developed a future-oriented strategic plan, which allowed actors to find solutions that were grounded in the local context and time-flexible. A similar approach was initially planned to be used in Bodø. As a contradiction, in Turku, the lack of a future mobility-oriented strategy was demonstrated as a prioritization of other selected initiatives, although internal development toward future mobility was considered.

In summary, providing a systemic change to adopt aerial services necessitates anticipatory consideration when building new area development initiatives to ensure a smooth transition. This change is time-consuming and multi-disciplinary, necessitating solutions for both airborne services and infrastructures that are flexible, adaptable, and capable of future transformation.

5 Discussion

5.1 New Mobilities Impacting City Development

Emerging forms of mobility are introducing complex and often unpredictable effects on urban development. Among these, electric and renewable energy-powered vehicles offer promising pathways toward more sustainable transportation systems for both goods and people. In particular, aerial mobility solutions present opportunities to establish new types of urban connections that are not constrained by existing road infrastructure, potentially reshaping the spatial logic of cities.

However, the introduction of novel transportation modes brings both opportunities and challenges at the systemic level. These innovations can generate significant socioeconomic impacts, including the displacement of certain job categories and the creation of new employment opportunities in sectors such as services and technology for AAS. Addressing these shifts requires a holistic approach to mobility infrastructure—one that situates mobility hubs within broader societal and economic systems.

In practical terms, the design and development of mobility hubs should not only respond to current user needs and service demands but also anticipate future transformations. Spatial planning must incorporate flexibility and adaptability, enabling these hubs to evolve alongside changing mobility patterns and technological advancements. In doing so, mobility hubs can serve as strategic instruments for guiding sociotechnical transitions, supporting inclusive economic development, and ensuring resilience in the face of future uncertainties.

5.2 Infrastructure Supporting Various Autonomous Services in Urban Environments and Vice Versa

Built-environment solutions, such as mobility hubs, are a key driver of the transition to attractive autonomous services. In Europe, they are increasingly used to meet growing mobility demands, as shown in the three case studies discussed. The potential of using innovative forms of mobility, such as AAS, can respond to the development needs in Europe and enable new ways to implement interconnected transport of goods and people in urban environments. The design and implementation of mobility hubs may take place as a common approach across Europe, which highlights the opportunity for a wide impact on the adoption of future autonomous aerial services and the implementation of the EU Drone Strategy 2.0.

Mobility hubs provide one example of the infrastructure that supports autonomous services, whether aerial or land-based. These numerous autonomous services facilitate logistics, urban maintenance, other techniques of manipulating and reacting to the urban environment, and data collecting in the city, all of which have an influence on the requirement for infrastructure. Infrastructure responses can range from a basic movable streetscape intervention to a large-scale fixed infrastructure facility, depending on the required service functionality.

The construction of infrastructure presents the challenge of how to best deploy it from a project standpoint. The chosen project organization model influences the ability to provide transition-enabling flexible infrastructure within specific projects. For example, to respond to growing mobilities with mobility hubs, projects must incorporate additional work and include components for now-unknown activities, such as fully established airborne services that are not yet important during development. The necessity for flexible solutions causes uncertainty in the building process, which can be detrimental to some project stakeholders, resulting in internal conflicts. In general, there exists a trend to act on present interests rather than foresight or anticipation when it comes to urban development. The concept of flexibility for future needs presents challenges for project organization, but it also enables future-proof opportunities for general urban development, which can be argued to be more sustainable over time as the need for resource-intensive renewals or modifications is reduced. Furthermore, to motivate possible development, initiatives can play a pioneering role in the emerging aerial service transition and have a significant impact on the future.

5.3 Question of Control and Ownership for Autonomous Operations in the City Regions

Implementing autonomous services, whether airborne or ground-based, raises the question of who controls and operates the city's autonomous robots and services. While liability remains a central concern when autonomous systems cause harm or disruption, broader systemic challenges also emerge. One such issue involves the tension between public ownership of urban infrastructure and the encroachment of privately operated autonomous services. For instance, food delivery robots navigating pedestrian walkways and road networks may conflict with public expectations for accessible and unobstructed public space. These services can also overlap with existing transportation or delivery systems, contributing to congestion and diminishing the quality and availability of shared urban spaces.

Furthermore, private independent service operators select where they wish to operate, which can result in unequal access across neighborhoods or cities, raising concerns about fairness and discrimination, as seen in the United States [57]. Additionally, these robots often collect large amounts of data, creating privacy risks and raising questions about who owns and controls data gathered in public spaces.

In response to these challenges, some cities have begun exploring regulatory frameworks to assert democratic control over urban airspace. For example, several European municipalities have proposed zoning strategies for aerial autonomy, aiming to retain local authority over airspace management rather than deferring entirely to national aviation regulators. These zoning initiatives not only reflect a desire to safeguard municipal sovereignty but also have practical implications for the integration of autonomous aerial systems (AAS) into urban mobility infrastructures. Specifically, airspace zoning may influence the design and functionality of mobility hubs, which serve as key nodes for multimodal transportation exchange. Interviews conducted in Finland and Norway reveal that discussions around municipal-level airspace governance are already underway, indicating a growing interest in localized regulatory approaches in Northern Europe.

Autonomous mobility-based services might be viewed as a catalyst for increased service efficiency and further automation in urban areas. Autonomous robots support the formation of a connected fleet of vehicles, while robotic services enable the materialization of AI into physical space via interaction in real space. For autonomous robotic services, such as intelligent transportation and the maintenance of the built environment, the collaborative advances of machine learning and robotics into physical space are a prerequisite for the safe and inclusive adoption of robotic services. However, their integration must be approached holistically, considering impacts on public space, infrastructure, and inclusive urban development.

6 Conclusion

The study highlights the transformative potential of autonomous aerial systems (AAS) in promoting sustainable urban and regional development, particularly in remote and constrained areas like the Arctic. By analyzing case studies in Berlin, Bodø, and Turku, we have provided prescriptive guidelines for urban development projects to enable the transition to these systems. Central to the prescriptive advice of the research is the establishment of mobility hubs, envisioned as multifunctional, future-proof infrastructures capable of accommodating emerging mobility needs, including aerial services. These hubs demonstrate varying levels of preparedness and strategic alignment across the examined cases while maintaining cost-effective approaches for temporary uses prior to the maturity and adoption of aerial services.

The findings emphasize the necessity of incorporating flexible, adaptive designs in urban planning to support the seamless integration of autonomous systems over time. For instance, the Berlin TXL project exemplifies a proactive approach to mobility innovation, leveraging a future-oriented development strategy to align future-proof infrastructure with emerging technologies. Similarly, Bodø's development benefits from its aviation-centric culture, aiming to enhance regional connectivity and resilience through AAS. Turku's focus on incremental improvements in mobility hubs signals a need for greater strategic foresight in anticipating future technological shifts.

Key enablers for successful AAS integration include leveraging local contexts and unique regional characteristics, fostering temporal flexibility in infrastructure, and embedding forward-looking strategic drivers into development projects. The study underscores balancing innovation with regulatory and operational challenges, such as airspace zoning and interoperability with existing transport systems.

In conclusion, the transition to autonomous aerial systems demands a holistic, anticipatory approach that aligns the development of infrastructure, policy, and technology. By addressing these elements cohesively, urban and regional planning can harness the potential of AAS to foster sustainable, resilient, and connected communities, thus contributing meaningfully to global sustainability goals. This research offers actionable insights and structured guidance for policymakers, planners, and stakeholders in navigating the complexities of integrating AAS into the future urban landscape.

References

1. Leminen, S., Rajahonka, M., Wendelin, R., Westerlund, M., & Nyström, A.-G. (2022). Autonomous vehicle solutions and their digital servitization business models. *Technological Forecasting and Social Change, 185*, 122070.
2. Clement, J., Ruysschaert, B., & Crutzen, N. (2023). Smart city strategies—A driver for the localization of the sustainable development goals? *Ecological Economics, 213*, 107941. https://doi.org/10.1016/j.ecolecon.2023.107941
3. Cohen, A. P., Shaheen, S. A., & Farrar, E. M. (2021). Urban air mobility: History, ecosystem, market potential, and challenges. *IEEE Transactions on Intelligent Transportation Systems, 22*, 6074–6087. https://doi.org/10.1109/TITS.2021.3082767
4. Geels, F. W., & Schot, J. (2007). Typology of sociotechnical transition pathways. *Research Policy, 36*, 399–417.
5. Kaartemo, V., & Helkkula, A. (2018). A systematic review of artificial intelligence and robots in value co-creation: Current status and future research avenues. *Journal of Creating Value, 4*, 211–228. https://doi.org/10.1177/2394964318805625
6. Samouh, F., Gluza, V., Djavadian, S., Meshkani, S., & Farooq, B. (2020). Multimodal autonomous last-mile delivery system design and application. In *2020 IEEE International Smart Cities Conference (ISC2)* (pp. 1–7). https://doi.org/10.1109/ISC251055.2020.9239082.
7. Geels, F. W. (2002). Understanding the dynamics of technological transitions. A co-evolutionary and socio-technical analysis
8. Geels, F. W. (2012). A socio-technical analysis of low-carbon transitions: Introducing the multi-level perspective into transport studies. *Journal of Transport Geography, 24*, 471–482. https://doi.org/10.1016/j.jtrangeo.2012.01.021
9. Geels, F. W. (2005). The dynamics of transitions in socio-technical systems: A multi-level analysis of the transition pathway from horse-drawn carriages to automobiles (1860–1930). *Technology Analysis & Strategic Management, 17*, 445–476.
10. Pons-Prats, J., Živojinović, T., & Kuljanin, J. (2022). On the understanding of the current status of urban air mobility development and its future prospects: Commuting in a flying vehicle as a new paradigm. *Transportation Research Part E: Logistics and Transportation Review, 166*, 102868.
11. Darling, K.: The new breed: How to think about robots. Penguin (2021).
12. Saravanakumar, Y. N., Sultan, M. T. H., Shahar, F. S., Giernacki, W., Łukaszewicz, A., Nowakowski, M., Holovatyy, A., & Stępień, S. (2023). Power sources for unmanned aerial vehicles: A state-of-the art. *Applied Sciences, 13*, 11932. https://doi.org/10.3390/app132111932
13. European Commission. (2021). *A Drone Strategy 2.0 for Europe*. https://ec.europa.eu/info/law/better-regulation/have-your-say/initiatives/13046-A-Drone-strategy-20-for-Europe-to-foster-sustainable-and-smart-mobility_en
14. EASA. *U-Space: A quick overview*. https://www.easa.europa.eu/en/what-u-space. Last accessed June 6, 2025.
15. Commission Implementing Regulation (EU) 2019/947 of 24 May 2019 on the rules and procedures for the operation of unmanned aircraft (Text with EEA relevance) (2019).
16. Forum Virium, S. CITYAM empowers responsible urban air mobility. The CITYAM project aims to support and empower cities in facilitating a responsible and acceptable increase in urban air mobility via transnational Baltic Sea Region cooperation, in order to achieve a cleaner and more sustainable transportation system. Last accessed November 26, 2024.
17. Torabi, F. K., Araghi, Y., van Oort, N., & Hoogendoorn, S. (2022). Passengers preferences for using emerging modes as first/last mile transport to and from a multimodal hub case study Delft campus railway station. *Case Studies on Transport Policy, 10*, 300–314. https://doi.org/10.1016/j.cstp.2021.12.011

18. Aminzadeh, A., Dimitrova, M., Meiabadi, M. S., Sattarpanah Karganroudi, S., Taheri, H., Ibrahim, H., & Wen, Y. (2023). Non-contact inspection methods for wind turbine blade maintenance: Techno–economic review of techniques for integration with Industry 4.0. *Journal of Nondestructive Evaluation, 42*, 54.
19. Gehrke, S. R., Phair, C. D., Russo, B. J., & Smaglik, E. J. (2023). Observed sidewalk autonomous delivery robot interactions with pedestrians and bicyclists. *Transportation Research Interdisciplinary Perspectives, 18*, 100789. https://doi.org/10.1016/j.trip.2023.100789
20. Lemardelé, C., Estrada, M., Pagès, L., & Bachofner, M. (2021). Potentialities of drones and ground autonomous delivery devices for last-mile logistics. *Transportation Research. Part E, Logistics and Transportation Review, 149*, 102325. https://doi.org/10.1016/j.tre.2021.102325
21. Legacy, C., Ashmore, D., Scheurer, J., Stone, J., & Curtis, C. (2019). Planning the driverless city. *Transport Reviews, 39*, 84–102. https://doi.org/10.1080/01441647.2018.1466835
22. Rajendran, S., & Shulman, J. (2020). Study of emerging air taxi network operation using discrete-event systems simulation approach. *Journal of Air Transport Management., 87*, 101857. https://doi.org/10.1016/j.jairtraman.2020.101857
23. United States Architecture News. Foster + Partners Releases Design For Uber Air Skyport in Santa Clara. https://worldarchitecture.org/architecture-news/ececf/foster%2D%2Dpartners-releases-design-for-uber-air-skyport-in-santa-clara.html. Last accessed November 26, 2024.
24. Skyport Infrastructures, https://skyports.net.
25. Fairs, M.. *London rooftops snapped up for "vertiports" as drone travel moves closer*. https://www.dezeen.com/2018/08/23/skyports-barr-gazetas-london-rooftops-vertiports-drones-technology/. Last accessed November 26, 2024.
26. Curlander, J.C., Gilboa-Amir, A., Kisser, L.M., Koch, R.A., & Welsh, R.D.. (2017). *Multi-level fulfillment center for unmanned aerial vehicles*. https://patents.google.com/patent/US20170175413A1/en
27. Berg, P.W., Isaacs, S., & Blodgett, K.L.. (2016). *Airborne fulfillment center utilizing unmanned aerial vehicles for item delivery*. https://patents.google.com/patent/US9305280B1/en
28. Matternet. https://www.matternet.com
29. Zipline. https://www.flyzipline.com
30. Amazon, S.. *Amazon drones can now fly farther and deliver to more customers following FAA approval*. https://www.aboutamazon.com/news/transportation/amazon-drone-prime-air-expanded-delivery-faa-approval. Last accessed November 26, 2024.
31. Wing, A.. *Wing Company site*. https://wing.com/fi_fi/. Last accessed March 07, 2023.
32. Atherton, K.D.. *Amazon's delivery drone hive patent is an urban planning nightmare*. https://www.popsci.com/amazons-patented-drone-towers-dont-make-any-sense/. Last accessed November 26, 2024.
33. Leffler, B.. *Vision für das ICC: Ein Zentrum für E-Mobilität in der Berliner City West*. https://entwicklungsstadt.de/vision-fuer-das-icc-ein-zentrum-fuer-e-mobilitaet-in-der-berliner-city-west/. Last accessed Jube 12, 2025.
34. Tiddi, I., Bastianelli, E., Daga, E., d'Aquin, M., & Motta, E. (2020). Robot–city interaction: Mapping the research landscape—A survey of the interactions between robots and modern cities. *International Journal of Social Robotics, 12*, 299–324. https://doi.org/10.1007/s12369-019-00534-x
35. Cugurullo, F. (2020). Urban artificial intelligence: From automation to autonomy in the Smart City. *Frontiers in Sustainable Cities, 2*. https://doi.org/10.3389/frsc.2020.00038
36. Toyota Woven City|TOP|What is Woven City. https://www.woven-city.global/. Last accessed November 27, 2024.

37. Pangyo to Gwanghwamun in 15 Minutes. *Seoul's UAM Era begins with test flights along the Hangang River Next Year*. https://english.seoul.go.kr/pangyo-to-gwanghwamun-in-15-minutes-seouls-uam-era-begins-with-test-flights-along-the-hangang-river-next-year/. Last accessed November 27, 2024.
38. Craig, T. *Skyports & RTA start work on Dubai vertiportSkyports Infrastructure*. https://www.airportsinternational.com/article/skyports-rta-start-work-dubai-vertiport. Last accessed November 26, 2024.
39. FF2020, AiRMOUR, AURORA Workshop 14.12.2022—FF2020 presentation.pptm, https://drive.google.com/file/d/1PzFIiAlzqKOIlqZhbcd9UuEWdnKytzej/view?usp=share_link&usp=embed_facebook. Last accessed March 9, 2023.
40. Tegel Projekt GMBH. *Berlin TXL—Tegel Projekt GMBH*. https://tegelprojekt.de/en/press-information/. Last accessed February 8, 2023.
41. Demonstrations. *Safir-Ready*. https://safir-ready.eu/demonstrations/. Last accessed November 27, 2024.
42. Edelman, H., Stenroos, J., Peña Queralta, J., Hästbacka, D., Oksanen, J., Westerlund, T., & Röning, J. (2023). Analysis of airport design for introducing infrastructure for autonomous drones. *Facilities, 41*, 85–100.
43. Tegel Projekt GMBH. *Berlin TXL—The Urban Tech Republic*. https://tegelprojekt.de/wp-content/uploads/2022/12/221123_Basistext_UTR_EN.pdf. Last February 8, 2023.
44. Tegel Projekt GMBH. *Berlin TXL—Schumacher Quartier*. https://tegelprojekt.de/wp-content/uploads/2022/12/221123_Basistext_SQ_EN.pdf. Last accessed February 8, 2023.
45. Stenroos, J. (2023). Design-driven innovation approach for future-proof air-ground infrastructure: Case Schumacher quartier.
46. Bodø Kommune. (2024). *Mål og retningslinjer for planlegging av ny bydel*. https://bodo.kommune.no/_f/i58c30b45-b68a-4073-acfe-24526d0ffe75/mal-og-retningslinjer-for-planlegging-av-ny-bydel.pdf
47. City of Turku, Lundén Architecture Company. (2024). *Turun Tiedepuiston visio*. https://www.turuntiedepuisto.fi/sites/default/files/atoms/files/turun_tiedepuiston_visio.pdf
48. Our overall concept. https://schumacher-quartier.de/en/our-overall-concept/. Last accessed November 27, 2024.
49. Tegel Projekt GmbH ed. *Urban development framework plan Schumacher Quartier*. https://schumacher-quartier.de/wp-content/uploads/2022/11/191016_SQ_Rahmenplan.pdf
50. Bodø Kommune. (2024). Mobilitetsplan, https://bodo.kommune.no/_f/if46546f8-8406-463c-8450-8f990ca03bda/mobilitetsplan.pdf
51. City of Turku, Lundén Architecture Company, Renell Käppi Arkkitehdit & Cobe. (2024). Kupittaan Kärjen Kumppanuushanke, Asemakaavan Viitesuunnitelma. https://www.turuntiedepuisto.fi/sites/default/files/atoms/files/turun_tiedepuiston_visio.pdf
52. Pajumäki, P. (2024). Interview on logistics development strategies at Kupittaa area.
53. Langfjord, R. (2024). Hagen: Interview on the themes of urban development connected with the transition to autonomous aerial services.
54. Bodø Kommune. Planbestemmelser. https://bodo.kommune.no/_f/ifa5ecc21-0f91-4969-b46e-d2051a3e4031/planbestemmelser.pdf
55. City of Turku. (2024). *Sustainable urban mobility plan draft*. https://kerrokantasi-api.turku.fi/v1/download/sectionfile/192/
56. Huttunen, M. (2019). The u-space concept. *Air and Space Law, 44*, 69.
57. Woo, J., Whittington, J., & Arkin, R. (2020). Urban robotics: Achieving autonomy in design and regulation of robots and cities. *Connecticut Law Review, 52*, 319.

Drone Fertilization Experiments and Spraying Outlook in Agriculture

Jere Kaivosoja[1(✉)], Ari Ronkainen[1], Panu Korhonen[1], and Victor Bloch[2]

[1] Natural Resources Institute Finland (LUKE), Plant Production Unit, Tampere, Finland
{jere.kaivosoja,ari.ronkainen,panu.korhonen}@luke.fi
[2] Volcani ARO (Agricultural Research Organization of Israel), Rishon LeZion, Israel
victor.bloch@volcani.agri.gov.il

Abstract. Working drones are slowly becoming available and practically allowed in EU. They can provide alternative technologies or even disruptive solutions for practical farming. Although the carrying capacity of drone technologies is often criticized, the rapid refilling or payload switching could provide radical efficiency. Our goal was to test that, and see the future possibilities of spraying campaigns, as the regulations will preferably allow that soon in test purposes. We purchased a DJI Agras T-16 drone, and after SORA (Specific Operations Risk Assessment) process, we carried out two precision farming tests with additional fertilizations. The spreading missions were intense, where the 20 l tank was emptied during a 2 min flight, and an instant refill was accomplished. The battery was replaced after about four refills. We managed to spread about 200 kg per hour with our non-optimized tests aiming to spread 100 kg/ha on average. The granules were not gentle for the drone propellers and the drone's operational performance was near to its limits. Emptying the liquid tank of the same drone will take around 3–4 min, so pesticide spraying with refilling station could provide great possibilities after it is allowed. For the spraying utilization, fitness of international practices need to be evaluated in European context, practical use case experimentations are needed and methodologies for drift risk evaluations are needed.

Keywords: Spraying · UAV · Aerial spraying · Aerial spreading

1 Introduction

Working drones represent an interesting frontier in agricultural technology, offering innovative solutions to long-standing challenges in crop management. As precision agriculture continues to evolve, drones are playing an increasingly vital role in enhancing the efficiency and sustainability of farming practices. Technologies for drone-based spraying and spreading are rapidly becoming general worldwide. By using unmanned aerial vehicles (UAVs) to deliver pesticides, fertilizers, seeds, and other essential treatments, working drones allow for more accurate application, minimizing environmental impact and offering disruptive technologies for agriculture. OECD report [1] presented the current state of the art of UAV spraying systems in agriculture and [2] comprehensively

M. Laakkonen (ed.), *DroneAI for Resilient Nature*,
https://doi.org/10.1007/978-3-032-19008-6_3

reviews the technology and operation processes. As the agricultural industry seeks to meet the growing global demand for food while minimizing ecological footprints, applications into drone spraying and spreading offer promising avenues for achieving sustainable and efficient farming practices. These spraying drone applications has especially potential to provide benefits such as reduction of applicator exposure and the precise spot application contributing to the more sustainable use of pesticides [3, 4].

Drone-based spraying and spreading approaches are typically criticized by the low payload capacity. Typical agricultural spreading or spraying application rate is over 100 kg/ha, and that sets a huge challenge for drone technologies in arable farming. However, if the refilling is fluent, this limitation can be overcome.

Drone-based pesticide spraying is currently prohibited in Finland, however the fertilizer spreading is allowed within permission and relevant SORA (Specific Operations Risk Assessment) procedure. It is obvious that these drone technologies could offer sufficient solutions for small scale, replacing for example backpack spraying in horticulture applications but it would be interesting to find out the real performance limitations in agricultural fields and find out the challenges that needs to be met before the pesticide sprayings can be performed. Our hypothesis is that by testing the practical performance of fertilizer spreading, we can calculate theoretical potentials with spreading and spraying operations. Our research questions are: (1) what are the practical performance limits in drone spraying and spreading in agriculture? (2) Which kind of national practices are needed to enable aerial pesticide spraying?

2 Material and Methods

Our research had three phases: first we reviewed literature on how the spraying and spreading drones have been used in relevant environment, what kind of equipment is available and what are the determined best practices. In the second phase, we accomplished drone spreading trials to determine practicalities. In the third and last phase, we studied deeply the status of spraying regulations to determine further development needs to enable pesticide spraying.

2.1 Equipment and Best Practices

Best management practices for safe and effective application of pesticide using unmanned aerial spray systems [5] compiles the resent relevant general issues. It summarizes current regulations in key markets, including examples of licensing requirements, presents the user safety in the context, and suggests equipment set up and calibration parameters that impact spray deposition.

As the refilling and battery replacement can be a straightforward procedure, the main factor in the spreading and spraying missions is the spraying and spreading speed. Typically, granule tank can be emptied within tens of seconds so the spreading pattern with the maximum speed is important, but with spraying the pump capabilities are the key.

We studied different kind of commercial spraying drones and listed them to following Table 1. Since new models are constantly released, our goal was not to seek complete list of models but to find example solutions from different manufacturers.

Table 1 Table captions should be placed above the tables

Brand	Model	Tank volume (l)	Spray flow/min (l)	Tank drain time (min)
DJI	Agras T40 [6]	40	12.0	3.3
DJI	Agras T16 [7]	16	4.8	3.3
Draganfly	Precision spray drone [8]	10	3.0	3.3
Hongfeidrone	HF T92 [9]	92	24.0	3.8
HSE-UAV	TTA M4E [10]	5	2.2	2.3
Hylio	AG-272 [11]	68	18.9	3.6
XAG	P100 [12]	40	6.0	6.7

Other spraying drone brands available and used in research publications are such as [2] 3DXR E610P and 3WYD-4-22A.

As it can be seen from Table 1, with full speed the tank will typically be empty after a bit over 3 min. There are also drones with petrol engine, such approach provides long operation times which is not needed in missions with constant refilling of liquids or granules.

The efficiency of the spraying or spreading depends greatly on the flying parameter optimization, refilling possibilities and battery management. The tank should last a single back and forth path flight. The spreading coverage is a result of the optimization of the flying altitude and swath width, speed of the spreader or sprayer pump and the outlet size related to the granule size and weight (see Fig. 1).

As the tank filling should be rapid, it should be predesigned. Applying submersible pump to nursing tank should allow 100–150 l/min refilling speed. Another important issue is the needed number of people in the operation. Regulations in Finland does not set requirements so theoretically a single person could manage the operation. However, the workload is rather high requiring high concentration all the time. If two persons are operating, one is managing the drone operation and other is taking care of the refilling and battery changes. Especially if battery charging and liquid mixtures are required, the tasks become too complicated for a single person to manage safely.

According to Garcia-Munguia et al. [2], spraying drone missions have reached reduction of 30% compared to the recommended typical dosage. This makes the applications already relatively advantageous.

2.2 Spreading Trials

We purchased and tested DJI Agras T-16 spreading drone (see Fig. 2) for fertilizer spraying missions in Jokioinen and Maaninka, Finland. This drone is relatively small, weighing just under 25 kg when operating without a payload. The operational flying time is about 15 min, and battery charging can be done in 20 min with 20 A power. Two persons were taking care of the missions.

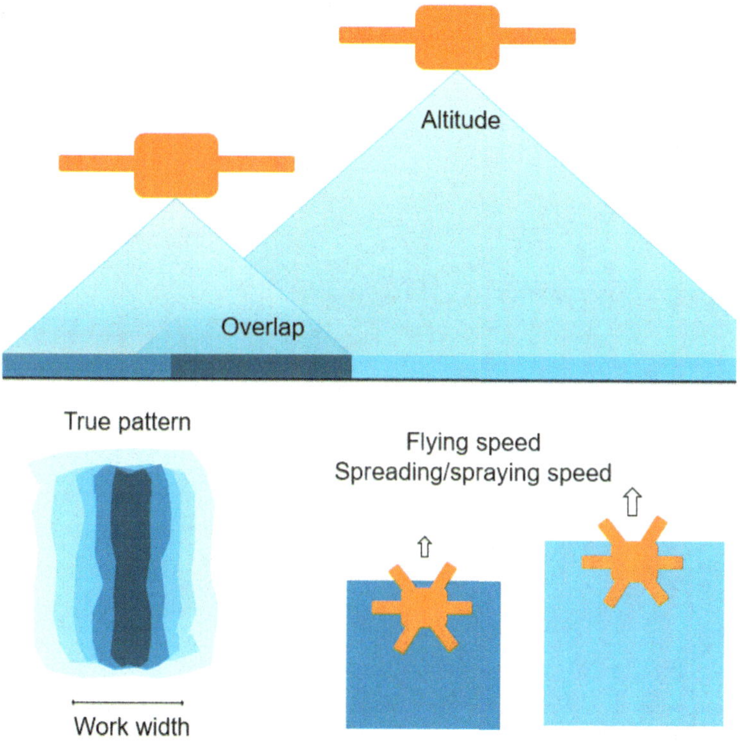

Fig. 1 Elements that effect on the true work result of the drone-based spraying and spreading

We made fertilization trial for the cultivated field by spreading YaraBela (Suomensalpietari, 27-0-1 NPK) granules. After preliminary tests, we aimed to spread 100 kg to a 1 ha area without any optimization. Granules were loaded to separate buckets used for filling the granule container on the drone. This mission took 30 min of time, required five refills and one battery replacement. One operational flight took about 2 min, meaning that of the 30 min mission, the drone was executing a spreading task actively for 10 min. Rest of the time went for transitions and for manual refilling. The spreader speed was set to 1000 rpm and the outlet size was 40%.

During the missions, the drone managed to spread fertilizers for few swaths, before it needed to return for the refill (see Fig. 3).

In Maaninka, the field was fertilized using drone on 25th June 2024, approximately three weeks after the first grassfield harvest. Before the experiment, the field was fertilized and managed following normal practices for grasslands containing clover. No additional fertilization was used after the first cut in drone trial area. Multispectral imaging of the area was done using DJI Phantom 4 Multispectral before the trial (24th June) and three weeks after the trial (16th July, see Fig. 4) to see if spreading lines are observable as plant indices (NDRE, Normalized Difference Red Edge).

Fig. 2 DJI Agras T-16 drone with the spreading unit. In addition, spraying nozzles are attached to the axis. Prefilled fertilizer granule buckets on the left, drone filling on the right

Fig. 4 NDRE-map processed in Solvi.ag (imaged 16.7.2024). N fertilized strips can be observed as a change in NDRE which is often used for N stress assessments

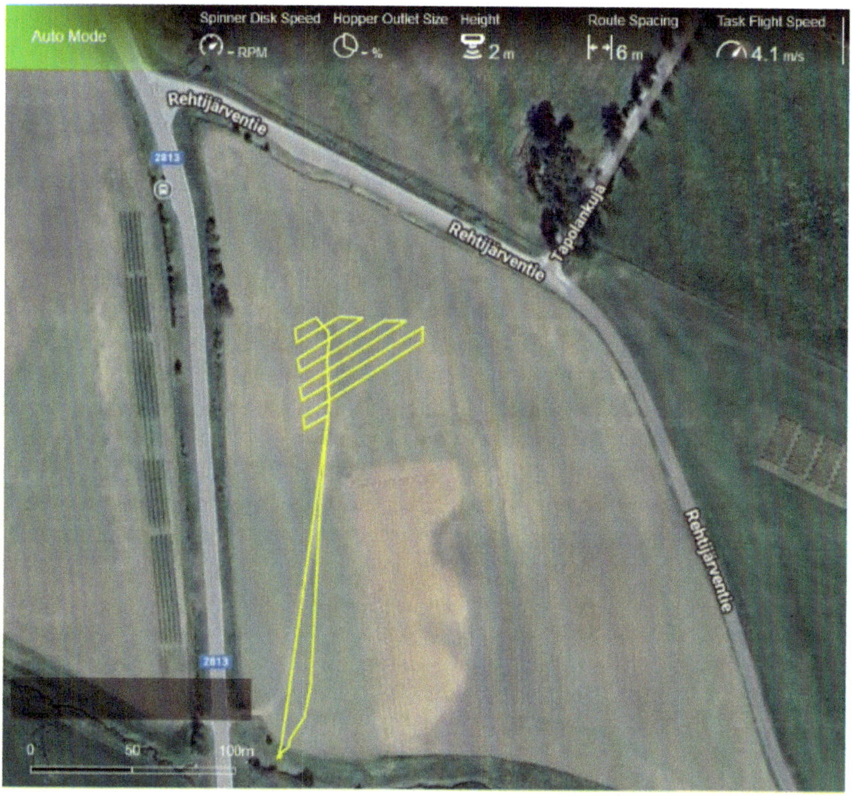

Fig. 3 A single flight including back and forth flights to the filling station, DJI Agriculture application

2.3 Toward Pesticide Spraying Practices

As it stands, it is generally not possible to apply pesticides in the European Union (EU) as the current legislation prohibits the aerial spraying to which the unmanned aerial spray systems (UASS) are included [13]. There are possibilities to use UASS in certain EU Member States (MS) based on a national derogation, but these are limited use cases in limited geographical areas [13, 14].

To overcome the legal constraint a change in legislation is needed. Means to estimate and manage risks related to application of plant protection products (PPP) from UASS of are needed to support change in legislation and to enable the possibility to use plant protection products from UASS. The authorities must be able to estimate the risks related to the pesticide application method, when using certain type of pesticide application equipment (PAE), in this case an UASS and the related risks of the PPP in this application. Authorities need to evaluate risks related to drift, bystander and operator risks, residue risks and efficacy [15].

One of the factors in risk assessment is the spray drift. Spray drift is a combination of several factors, such as drone size, configuration or type, spray altitude, speed, flowrate and nozzles, or spray system, and external factors such as wind [16–18]. Experiments show that in some conditions the drift from a UASS can be as low or lower than from an air blast sprayer [19]. This is however still more than from a field crop sprayer. The use of UASS with higher drift is often justified with lower treatment areas. However, it remains a question, how much smaller the total treatment area has to be for the risk to be acceptable.

The risks for bystanders and operators need to be evaluated as well. The risk from drift exposure is similar to drift in general, but there are specific test methods for evaluation [19, 20]. In addition to the exposure risk the operator is also exposed to the risks arising from the handling of the products and equipment. There are risks related to mixing and loading of PPP to the UASS, changing and filling of the spray tanks [16, 17]. These are point source exposures that can affect the environment or the operator. The concentration and exposure amounts are usually very high in direct exposure. It needs to be considered that an UASS is handled differently than traditional sprayers and therefore related risks are different. Experiments show that the UASS gets contaminated with PPP in operations. Experiments show that the contamination of UASS surfaces can be three times as high as with air blast sprayers [18, 21]. The problem with surface contamination is that UASS are usually handled and transported manually by lifting and in transport equipment. This requires proper use of personal protective equipment (PPE) to avoid secondary transfer. Also, vehicles or containers used to transport the UASS, and related components, become contaminated.

For handling the risks arising from working with PPP best management practices (BMP) are needed. Unmanned Aerial Pesticide Application System Task Force (UAPASTF) has released such guide [16]. The BMP describes managing drone operations and related risks. These guidelines will still need adaptation to local conditions. Form managing drift risks information from UASS manufacturers are needed to evaluate a suitable operations and spraying envelope.

For the application method to be acceptable it also needs to be effective in the purpose. Efficacy is one aspect that needs to be considered when approving a plant protection product (PPP). Efficacy is combination of the substance penetration and disposition in the canopy, in a concentration, that is sufficient to accomplish the goal and low enough to avoid unnecessary or unacceptable plant damage and stress. Evaluation of the efficacy must be made together with the PAE and PPP as both are relevant. This requires experimentation. Experiments have shown that it is possible to find combinations, where stronger concentrations of PPP can be used without excess damage [17].

EU legislation also requires regular inspection of PAE, including UASS, regularly with three-to-five-year intervals. For traditional PEA there are harmonized standards that define the test methods and acceptance criterion [A1]. New PAE also needs to be designed so that they can be tested and calibrated [12]. No such requirements exist for UASS. ISO 23117-1 and future ISO 23117-2 will probably, in future, offer requirements and methods for environmental requirements and testing of thereof. UAPASTF BMP offers guidance for calibration of UASS [16], but it remains unclear to the authors if such calibration procedure can be performed with the most common UASS currently in the European market.

3 Results and Discussion

As results, we present our estimations of potential performance of spraying and spreading drones and future development needs for the spraying operations.

3.1 Potential Performance

We calculated potential values for different types of commercial drones related. According to our field tests, we evaluated that tank refilling and transition flights can be managed in 2 min time on average. It is worth to mention that in such way, filling up 96 liter tank require efficient pumps and thorough mission planning. The first column in Table 2 presents the amount of liters that the drone model can spray within an hour, when the tank filling pitstop takes 2 min of time and the mission takes exactly the time that it takes for the drone to pump the tank empty. The next columns are calculated as follow: (A) how many hours it takes for the drone to spray 5 ha of field with a 150 l/ha application rate? (B) How many pitstops are needed in case of spraying that 5 ha? (C) If spraying 100 l/ha, how big area the drone can cover in 2 h (ha)? D) During a relevant one-day operation (6 h), how big area the drone can cover with the application rate of 200 l/ha? According to these calculations, spraying drones can cover up to 10–20 ha per day with decent application rate.

Table 2 Limits and theoretical performance of different drones

Name	Liters/h	(A) 5 ha, 150 l/ha (h)	(B) Refills in A	(C) 2 h, 150 l/ha (ha)	(D) 6 h, 200 l/ha (ha)
DJI Agras T40	288	2.6	19	5.8	8.6
DJI Agras T16	115	6.5	47	2.3	3.5
Draganfly	72	10.4	75	1.4	2.2
HF T92	625	1.2	8	12.5	18.7
HSE-UAV TTA M4E	41	18.2	150	0.8	1.2
Hylio AG-272	475	1.6	11	9.5	14.2
XAG P100	206	3.6	19	4.1	6.2
Average	260	6.3	19	5.2	7.8

The application rate of 200 l/ha is quite common pesticide spraying rate with a tractor. Typical tractor-based application rates vary between 100–400 kg/ha with different applications. According to the results in (Table 2), small spraying drones capable of spraying less than 200 l per hour can efficiently treat less than 5 ha per day while the biggest ones reached nearly 20 ha. Also, the number of refills with smaller units were enormous. These values suggest, that DJI Agras T40 with 40 l tank and dual spraying system [6] or XAG P100 represents the minimum requirement in the applications where application rate is over 100 l/ha, and the fields are several hectares.

In all of the cases, the fluent refilling is essential. Commercially available trailers and refilling stations for working drones vary in functions and capacity, but typically they are built of a flight deck, generators for charging drone batteries, tanks for storing liquids or granules, plumbing, and pumps and hoses for filling in the drones on the upper deck. One example is a nuWay Ag Drone trailer [22] marketed in US which is a two-story construction available with generators, chargers, tanks (e.g., 4000 l tank) and pumps on the bottom deck and a flight deck with a pilot protection area on top deck. Larger trailers can be used to operate two or more drones simultaneously. Development and adoption of such drone management systems to improve workflows are required before wider efficient application can take place on farms.

The lifecycle of the drone technologies is much shorter than lifecycle of common farm machinery. The lifecycle of a typical drone battery is around 200–500 recharges. So with these specifications, a single 1000 € battery may cover 200 ha of spraying work meaning 5 €/ha costs. With our test setups, the granules were too rough for the drone propellers and tiny pieces were broken (Fig. 5). This could have been greatly limited with more suitable settings, but the consumption of 400 € propeller sets is obvious. The drone was close to it's operational limits during our tests. The torque was high and there were minor electrical problems at the end.

Fig. 5 Fertilizer granules proved too rough for the drone propellers—causing tiny pieces to chip off and damage the blades

3.2 Future Development

Pesticide application from UASS has been a practice used widely in the world, except in Europe. This has created a knowledge gap between Europe and rest of the world. There are established best practices in place around the world and these practices must be adapted to the needs in EU and its Member States. Fitness of these practices need to be evaluated in European context. Establishment of European or national best management practices allows estimation of operation and flight mission related risks. To be able to establish these BMPs practical use case experimentations are needed. These give insight on the future uses of UASS in Europe and set requirements on the equipment and supporting operations on plant protection operations.

In addition to practical experimentation information about drift risks are needed. Drone type, flight parameters and environmental conditions should always be recorded, and drift measured when experimentation is made and published with the experiment results. These will help in determining the drift risks and risks related to different mission profiles, drone types and environmental conditions. Also, meta-analysis about drift related research is needed. At the moment, there might not be enough to perform meta-analysis. In the longer term it might be beneficial to move more toward physics-based modeling of drift risks such as using computational fluid dynamics (CFD) but developing computational tools might still take some time. Experimentation with drift should help to determine suitable flight or operations envelope for use of PPP from drones.

When using active agents, or actual PPPs, in experiments, there should always be a goal and reason for using actual PPP. This could be for evaluation of efficacy or usability in certain use cases. If the goal is to experiment solely with spray parameters or running of the operation, low risk marker substances should be used.

4 Conclusions

In a single day, 10–20 ha can be treated in a typical way, 200 kg/ha. This would require 50–100 refillings, so such work is intensively focusing on the fluent and safe refilling. Such operation would also require about 15–30 battery changes and failures related to high temperature could occur with constant recharging and flying. This is completely different if compared to typical drone missions or typical tractor missions. In such cases, an automatic filling station is needed. So, drone spraying and spreading as such are possible even in large field scale, but as the logistics and refilling automation are needed, the costs increase greatly above the single drone costs. Based on these aspects, drone-based spraying or spreading missions in bulk agriculture in large scale in Finnish conditions does not have a high potential to be cost beneficial. The beneficial cases should be found from high value crops, precision farming missions, greater active substance portions in liquid, or co-operative machinery, where for example the drone takes care of only the most difficult parts of the field.

For the spraying utilization in future, international best practices need to be evaluated in European context, practical use case experimentations are needed and research and measurements for drift risks are required. This work presented practical applicability of current drone technologies. From the scientific side, this work presented that these drone technologies can be implemented in large scale even with bulk agricultural products. Future research and development is needed in precision agricultural practices, granule spreading mechanisms and the integration of drone works as disruptive technology in agricultural production.

References

1. OECD. (2021). *Report on the state of the knowledge—Literature review on unmanned aerial spray systems in agriculture* (OECD series on pesticides, No. 105). OECD Publishing.

2. García-Munguía, A., Guerra-Ávila, P. L., Islas-Ojeda, E., Flores-Sánchez, J. L., Vázquez-Martínez, O., García-Munguía, A. M., & García-Munguía, O. (2024). A review of drone technology and operation processes in agricultural crop spraying. *Drones, 8*(11), 674. https://doi.org/10.3390/drones8110674
3. Berner, B., Moitzi, G., Pachuta, A., Maheswari, N., Wagentristl, H., & Chojnacki, J. (2019). Oil seed rape plant spraying using multirotor drone. In: *Farm machinery and processes management in sustainable agriculture.*
4. Wen, S., Zhang, Q., Yin, X., Lan, Y., Zhang, J., & Ge, Y. (2019). Design of plant protection UAV variable spray system based on neural networks. *Sensors, 19*(5), 1112. https://doi.org/10.3390/s19051112
5. Unmanned Aerial Pesticide Application System Task Force (UAPASTF). 2024. *Best management practices for safe and effective application of pesticides using unmanned aerial spray systems (UASS).* https://uapastf.com/wp-content/uploads/2024/09/MASTER-UAPASTF-BMP-final-Sept-2024.pdf
6. DJI Homepage. https://www.dji.com/fi/t40. Last accessed November 29, 2024.
7. DJI Homepage, https://www.dji.com/fi/t16. Last accessed November 29, 2024.
8. Dragenfly Homepage. https://draganfly.com/spraying-solution/. Last accessed November 29, 2024.
9. Hongfeidrones Homepage. https://www.hongfeidrone.com/hf-t92-agricultural-drone-92-liter-8-axis-foldable-transport-aircraft-product/. Last accessed November 29, 2024.
10. HSE-UAV Homepage. https://hse-uav.com/products/m4e-spraying-drone. Last accessed November 29, 2024.
11. HYLio Homepage. https://www.hyl.io/ag-272-product-page. Last accessed November 29, 2024.
12. XA Homepage. https://www.xa.com/en/p100. Last accessed November 29, 2024.
13. The European Commission. DIRECTIVE 2009/128/EC of the European Parliament and of the Council of 21 October 2009 establishing a framework for Community action to achieve the sustainable use of pesticides. Official Journal of the European Union, L 309/71.
14. OECD. (2021). *Report on the state of the knowledge—Literature review on unmanned aerial spray systems in agriculture* (OECD series on pesticides, No. 105). OECD Publishing.
15. Ronkainen, A., Kaivosoja, J., Laitinen, P., & Laitila, E.. (2022). Overcoming legislative requirements for implementing UAV plant protection operations in Europe. In: The XX CIGR World Congress 2022, Kyoto, Japan.
16. Unmanned Aerial Pesticide Application System Task Force (UAPASTF). Best management practices for safe and effective application of pesticides using Unmanned Aerial Spray Systems (UASS), September 20, 2024.
17. CropLife America Drones Working Group. (2021). UAV pesticide application: Benefits and fit into the current regulatory framework. *Industry whitepaper.*
18. Australian Pesticides and Veterinary Medicines Authority. (2021). State of the knowledge literature review on unmanned aerial spray systems. *Agriculture.*
19. Herbst, A., Bonds, J., Wang, Z., Zeng, A., He, X., & Goff, P. (2020). The influence of unmanned agricultural aircraft system design on spray drift. *Journal für Kulturpflanzen, 72,* 1–11.
20. A8European Food Safety Authority (EFSA), Charistou, A., Coja, T., Craig, P., Hamey, P., Martin, S., Sanvido, O., Chiusolo, A., Colas, M., & Istace, F. (2022). Guidance on the assessment of exposure of operators, workers, residents and bystanders in risk assessment of plant protection products. *EFESA Journal, 20*(1), 1831–4732.

21. Liu, Y., Li, L., Liu, Y., He, X., Song, J., Zeng, A., & Wang, Z. (2020). Assessment of spray deposition and losses in an apple orchard with an unmanned agricultural aircraft system in China. *American Society of Agricultural and Biological Engineers, 63*, 619–627.
22. NuWayAg Homepage. https://nuwayag.com/products/nuway-ag-drone-trailer. Last accessed November 29, 2024.

Enhancing Drone Operations During Wind Weather: Inertial Measurement Unit and Localized Weather Data

Rajeeka Hannes(✉), Markku Nuolimäki, Aleksi Kärnä, Henry Hinkula, Kimmo Paajanen, and Mika-Petri Laakkonen

Oulu University of Applied Sciences, Oulu, Finland
{rajeeka.hannes,markku.nuolimaki,aleksi.karna,henry.hinkula, kimmo.paajanen,mika-petri.laakkonen}@oamk.fi

Abstract. The use of Unmanned Aircraft Systems (UAS) has increased significantly in recreational, professional, and research and development environments. This has furthered the development of drones and increased the number of possible applications. In this article, we explore the utilization of Inertial Measurement Units (IMUs) in drones and their possibilities in measuring ongoing weather conditions compared to Automatic Weather Stations (AWS) data. In this research, test flights were conducted as part of our Arctic Drone Labs (ADLs) drone research center to gain comparable weather data and analyze the challenges related to using weather data by drones in Finland. Data is gathered from test flights conducted with UAS equipped with an IMU unit. Test flights were flown in the direct vicinity of local AWSs to maximize the comparability of weather data. Data analysis and comparison to AWS data revealed that weather data collected via onboard IMU sensors needs more test flights and further development to achieve more consistent data correlation. These results emphasize that IMU can be utilized to gather localized weather data, but more extensive applications require further development.

Keywords: Drones · Flight operations · Unmanned aerial vehicle · Data quality · IMU · Weather forecasting

1 Introduction

The popularity of Unmanned Aircraft Systems (UAS), more commonly known as drones, has been increasing in recent years [1]. More affordable prices have increased their availability, and increased use in various applications has brought new potential for research facilities and businesses. The utilization of UAS in various business and authority activities has increased significantly. UAS are available in different price ranges and equipped with varying features. UAS technology is utilized primarily in mapping applications, such as terrain modeling and photogrammetric aerial imaging. New developments in drone technology have also widened the use cases of UAS in search-and-rescue operations, animal monitoring in agriculture, delivery services, and safety applications [3, 10, 13].

M. Laakkonen (ed.), *DroneAI for Resilient Nature*,
https://doi.org/10.1007/978-3-032-19008-6_4

The Finnish Transport and Communications Agency (Traficom) controls Finland's drone regulations. In the regulatory context for drones, the generalized term UAS includes all systems that operate automatically, autonomously, or are remotely controlled [4]. The European Union Aviation Safety Agency (EASA) defines drones as unmanned aircraft. The European Union has a set of standard aviation rules and regulations for all member states. In addition, the Finnish government has legislation governing aviation rules, such as the Aviation Act and Aviation Regulations. These regulations facilitate development and safety for all users in aviation. Traficom is also responsible for keeping a registry of different drone operators, both private and commercial [4, 14, 15].

According to guidelines by Traficom [15], the drone operator must know the currently applicable rules and regulations regarding drones and their use. Flights using a drone while under fatigue, under the influence of alcohol or other intoxicating agents, or while otherwise impaired are prohibited. During drone operations, constant observation of other aviation vessels and continuous attention to other aviation safety factors are required [15]. Drones can be divided into multirotor, for example, DJI Mavic 3, and Nordic Drones Geodrone 6, and fixed-wing drones, for example, eVTOL Avy Area 4 QuantumSystems Trinity Pro. Multirotor drones can fly similarly to helicopters and are usually used in photography and videography applications. Fixed-winged drones resemble conventional aircraft and require higher air speeds to maintain lift and stay in flight. Higher air speeds make these more suitable for covering larger areas and longer distances [1, 3, 15].

The increased use of drones has highlighted thc importance of assessing current weather conditions during the operation. Reliable detection of flying conditions is a requirement for scalable drone applications [10]. Weather assessment and its related technologies generally lag behind the conventional aviation industry. Environmental factors like air temperature, wind, and humidity determine drone performance, durability, and maneuverability. Weather conditions assessment relies on static weather stations [7, 10].

Smart and sustainable aviation research group researches and develops the future of unmanned aviation and related industries. Inertial Measurement Unit (IMU) is a crucial drone sensor, acting as the drone's sensory system. It measures the drone's acceleration, rotation, and orientation, providing data essential for navigation and stabilization. From the experiences of its personnel stemmed an interest in the utilization of the Inertial Measurement Unit (IMU) in drones and its possibilities to measure and monitor weather conditions dynamically. IMU determines the drone's position, orientation, and movement, helping it navigate accurately and maintain stable flight. IMUs combine gyroscopes and accelerometers, which use these sensors to calculate the drone's attitude (roll, pitch, and yaw), velocity, and position. IMU (Inertial Measurement Unit) comprises three-axis accelerometers, gyroscopes, and magnetometers for global positioning [5].

An earlier research paper has developed techniques to improve heading alignment applied to the GNSS/INS integration scenario personalized for drone navigation problems using low-cost MEMs-IMU sensors. This research has also accounted for multidimensional parameter alignment, resulting in better accuracy of the overall navigation solution for the UAV platform. In addition, research presents actual operational flights to illustrate extension to the multivariable alignment, which is mathematically sound but still not evident [2].

The Oulu area was selected as an imaging area, and imaging took place in 2024. The focus of the research was the functionality of IMU during drone flight and its additional value for drone research. The primary goal was to acquire further experience using IMU technology and to gauge the accuracy of altitude and location data gathered by photogrammetric point clouds [3]. Data was collected from multiple flights flown with two drones, one DJI Mini 4 Pro and one Mavic 3 drone. Aerial imaging data was processed and analyzed. As a result, new information on the use of IM was generated, and based on data analysis, it can be concluded that further experiments with IMU in the use of drones are necessary for practical use. Aerial imaging and the use of IMU gave new insight and value as practical knowledge of the imaging process, and further research could focus on applying IMU to varying drone models and utilizing this technology in larger imaging areas in the Northern Ostrobothnia area.

2 Literature Review

The term 'drone' is often used to refer to any unmanned vehicle traversing through air, water or land autonomously or remotely controlled (Fig. 1). Compared to traditional methods, drones are an excellent fit for surveillance and surveying purposes thanks to their speed and cost-efficiency. Drones that are flown are usually referred to using abbreviations UAS, RPA, and RPAS [1]. UAS means Unmanned Aerial System, similar to RPAS, which is short for Remotely Piloted Aircraft System and includes control devices and software for piloting and tracking. RPA is the abbreviation for Remotely Piloted Aircraft, and in addition to smaller drones, includes remotely controlled unmanned aircraft and is also called UAV, Unmanned Aerial Vehicle [11, 12].

Fig. 1 Smart and sustainable aviation (SSA) research program's drones: (**a**) DJI Mini 4 Pro, (**b**) DJI Mavic 2, and (**c**) Geodrone

In recent years, light drones have become the remote sensing tool of choice among many experts, especially in ecology, nature preservation, and environmental sciences. In these fields, lightweight drones (Fig. 1a, b) have become an established part of remote sensing measurement methodology. Drone technology can be utilized in various tasks such as wildlife monitoring, plant life quantifying, biomass estimation, coastline morphological mapping, coral reef surveillance, natural disaster detection, and agriculture [14]. On the other hand, drone technology brings forth its issues as a lack of standard protocols can disturb collaboration between different parties, and gathered data might not be directly applicable outside their original field. Furthermore, this incentivizes different parties to purchase their equipment, reducing the cost-effectiveness. The form of gathered data can depend on location, user-dependent factors, and the methods used to gather the data. Data collected locally for individual parties often leads to a lack of scientific data reporting [3, 7, 8].

New practical applications are being searched for drones, enabling easy work in various types of terrain. Image data from drones can be classified as target, view, and panorama images. Oblique views, spherical panoramas, and video of the planned path can be used for visual modeling and documentation. Aerial imaging has a distinct impact on visual impressions compared to ground-level imaging. Laws and regulations regulate aircraft (EU 947/2019). New rules outline precise educational requirements, broader registration obligations, and scope for flight operations. Drones weighing under 25 kg must be registered, and a theory test must be passed to operate them. Also, the flight altitude cannot exceed 120 m in the open category. If the flight takes place at higher altitudes, it is classified into the special category. Operators flying in the special category must apply for an official permit and make a special announcement for each flight [10, 13, 16].

2.1 Drone Operations

Data gathering by drones is done through various sensors. The most common sensor is a camera capable of individual still images and video footage. Specialized multi- and hyper-spectrum cameras enable imaging outside the visible RGB –spectrum [9]. Infrared sensors are another type of special sensor which can be utilized to detect heat leaks in buildings. Various kinds of laser scanners can be utilized to produce accurate data for 3D modeling. The most common types of imaging data are still images and video footage. Oblique images and spherical panoramic images are most common in environmental surveillance. Imaging data can be combined using various software systems, such as Image Composite Editor, DJI Media Maker, or PTGui. FSP-Viewer, a freely available program, can view spherical panorama images. Mapping flights can be performed with many applications like Pix4D Capture, Pix4DMapper, Agisoft Metashape, WebODM, and DroneDeploy. Different applications provide different features and usability, but all can produce photogrammetric-based data such as georeferenced orthomosaic images, digital elevation models (Digital Surface Model), point clouds (LAS), or 3D surface models (Mesh) [16] (Fig. 2).

Ground Control Points (GCP) can be set up for accurate measurement of location in the imaging terrain, providing a reference point for location correction. Recently, it has become common for drones to have Real-Time Kinematic (RTK) technology, which does not require outside reference points. This technology can especially help with the rapid deployment and operation of individual drones. Flights are often flown on autopilot along a predetermined flight path, and the operator supervises the flight with

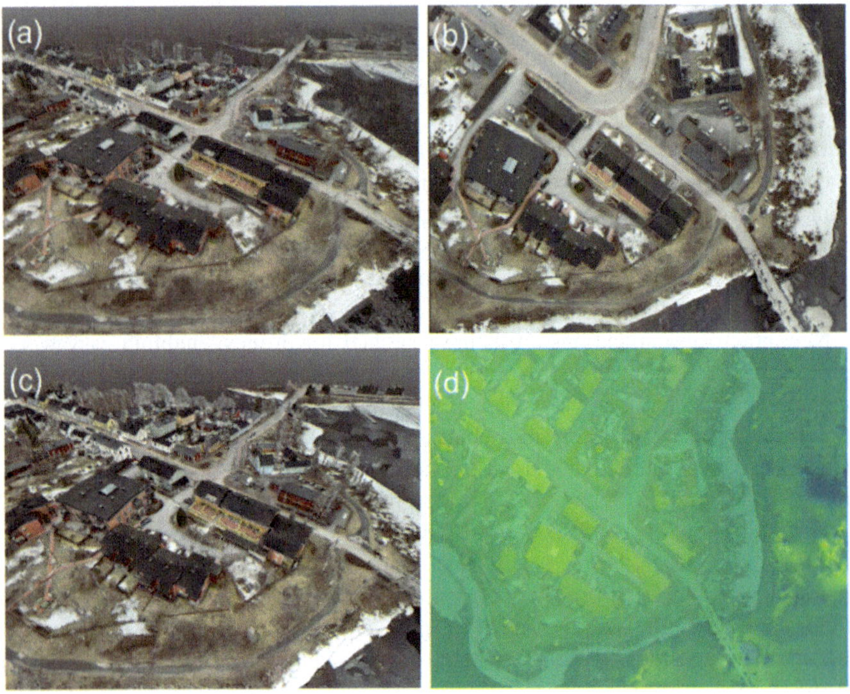

Fig. 2 Photogrammetric-based data models: (**a**) Point cloud, (**b**) Georeferenced orthomosaic, (**c**) Mesh, and (**d**) Digital Surface Model

visual contact. Take-off and landing are performed manually. Aerial aircraft imaging can produce a series of photos that can be transformed into an orthoimage mosaic [11]. Aerial imaging data can be computationally transformed into three-dimensional surface models such as DEM, DTM, and DSM. The Digital Elevation Model (DEM) is a model of terrain elevation levels, the Digital Terrain Model (DTM) maps the types of terrain, and the Digital Surface Model (DSM) models the shape of the surface [9, 11, 12].

Most commonly, drones are used for photography from different spectrums. Still, they can be used to measure the Earth's magnetic field and electric conductivity of the terrain, as well as gamma spectrometric measurements. Aircraft can also gather water samples and water quality surveillance, 3D modeling of mobile phone networks, and condition inspection of power lines and buildings. These applications require expertise, experienced pilots, and personnel to develop and produce the necessary software. Analysts are often needed to analyze and process the gathered data [8, 11, 12].

2.2 Weather Data Collection

At the Finnish Meteorological Institute, weather forecasts and weather statistics are essential in environmental research and producing weather and climate services. Weather information and observations are necessary for weather services for agriculture and forestry, seafaring, aviation and traffic, tourism, energy production, and maintenance. In

Finland, surface weather observations are made extensively, and most observations are gathered through automated weather observation stations. Weather observation stations are usually in open areas with relatively level terrain and no nearby obstacles. Observations are made in these stations once daily and in the mornings. In aviation, weather stations make observations every two hours. Modern stations are mostly automated; observations are received every 1–10 minutes. Basic metrics include air temperature, relative humidity, atmospheric pressure, wind, rainfall, and visibility. Weather observations are always recorded with a UTC timestamp (Universal Time Coordinated). When trying to understand atmospheric processes, one of the main parameters is wind speed and direction [6].

Wind observations over large spatial areas and precise time resolution are challenging. Automated weather stations are stationary near ground-level measurement points, and their data is limited. For this reason, unmanned aerial vehicles can provide a cost-effective and repeatable addition to weather data gathering. Drones have already been popular, especially in remote areas and harsh environments. Interest has also risen in wind energy production since drones provide flexible observations of vortexes [9, 10].

We used the DJI Mavic 3, which has a maximum ascent speed of 8 m/s, a descent speed of 6 m/s, and a horizontal speed of 21 m/s. It also has a maximum wind speed resistance of 12 m/s, a maximum pitch angle of 35°, and a global navigation satellite system of GPS + Gelileo + BeiDou.

The drone system demonstrates exceptional precision in both hovering and stabilization, ensuring reliable data collection under varying conditions. Vertical accuracy reaches ± 0.1 m with vision positioning and ± 0.5 m with satellite positioning, while horizontal accuracy ranges from ± 0.3 m to ± 0.5 m. Additionally, the 3-axis mechanical gimbal provides robust stabilization with minimal angular vibration, even in sport mode (see Table 1).

Table 1 Drone data collection accuracy and stabilization specifications

Drone data collection accuracy and stabilization specifications	
Hovering accuracy range	Vertical: ±0.1 m (with vision positioning) ±0.5 m (with satellite positioning) Horizontal: ±0.3 m (with vision positioning) ±0.5 m (with satellite positioning)
Stabilization	3-axis mechanical gimbal (tilt, roll, pan) Mechanical range: Tilt: −140° to 50° Roll: −50° to 50° Pan: −23° to 23° Controllable range: Tilt: −90° to 35° Pan: −5° to 5°
Max control speed (tilt)	100°/s

(continued)

Table 1 (*continued*)

Drone data collection accuracy and stabilization specifications	
Angular vibration range	Hovering without wind: ±0.001° Normal mode: ±0.003° Sport mode: ±0.005°

3 Experiments and Results

3.1 Comparison of an Automated Weather Station and Drones

In selecting the test site, we consulted with the Finnish Meteorological Institute and discussed various weather stations in the Oulu area and their measurement methods. The Oulu Hailuoto weather station was chosen as the most suitable for the test flights. The weather station is located in an open area, allowing for safe drone operations in its vicinity. As shown in Fig. 3, the weather station's location is marked on the map, along with the positions of the drones relative to the weather station.

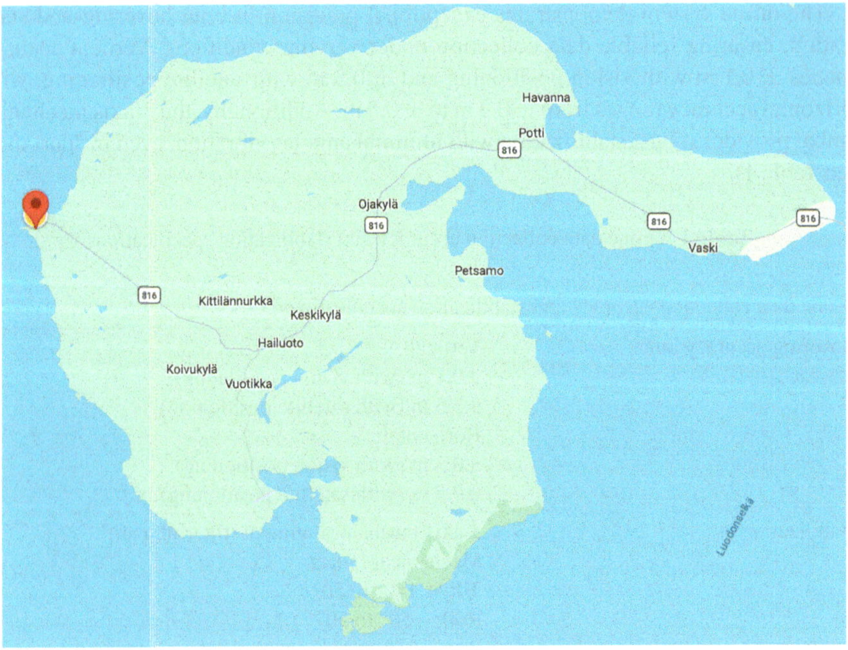

Fig. 3 The location of the AWS and the drone flight point are indicated by a red marker (65°02′23.1″N 24°33′42.1″E)

According to the plan, the flights aimed to maintain two drones at the same altitude as the weather station's sensor, approximately 40 m above sea level. Additionally, each drone was positioned at a specific location relative to the weather station. This approach allowed for measurements from multiple points while minimizing the impact of external structures or vegetation on the wind. The test flights were conducted on November 18, 2024, in Oulu Hailuoto, when wind conditions varied, providing an excellent test environment due to the changing weather.

The test flights were conducted from 08:10 to 12:20 (UTC). The flights were performed in pairs, followed by a break to recharge the batteries. Three pairs of flights were conducted throughout the day, resulting in six flights per drone, yielding 12 data measurements. During the flights, the drones calmly ascended and moved to the designated location near the weather station and at the specified altitude. Once the altitude was reached, the drones could hover using their GPS, and flight safety was monitored simultaneously. In a few instances, the drones drifted with the wind and had to be repositioned during the measurement, but this likely did not significantly affect the measured data or accuracy.

The equipment selected from Arctic Drone Labs for the test flights included the DJI Mini 4 Pro and a borrowed DJI Mavic 3. Since the drones are classified with C0 and C1 markings, the test flights could be conducted on short notice in the open category. A prerequisite for both drones was support in Airdata, allowing the log data to be easily converted into wind data. The flights were conducted within a visual line of sight from a quiet location, as shown in Fig. 3. The devices are consumer-grade drones priced under €1500, offering both affordability and performance.

Data from the drones was collected using internal sensors, including the IMU and compass, which were programmatically converted into wind data and direction. The collected log files were stored and transferred to the Airdata website for analysis, which promises direct wind data from the log files. Further details on data analysis are provided in Sect. 3.2.

Hailuoto Weather Station (Automatic Weather Station, AWS) is an automatic weather station managed by the Finnish Meteorological Institute. The weather station can measure weather conditions as well as air quality. The wind measurement is placed on the upper platform of the radar tower on a mast about 2 m long. The island affects the measurement accuracy for winds from the east (in the sector of 45°–165°). An acoustic sensor is used for wind measurement at the Hailuoto weather station. The sensor consists of three or four ultrasonic transducers that act as transmitters and receivers. When sound travels between the sensors, wind parameters can be calculated. Weather stations usually report hourly and 10-minute averages as well as three-second gusts. At the Hailuoto weather station, the measurement frequency is a one-minute average [6].

3.2 Data and Results

The personnel of Arctic Drone Labs are preparing for test flights in Hailuoto (Figs. 4 and 5). Data was gathered for this research from 12 test flights, six of which were flown with the DJI Mavic 3 drone, and the other six test flights were flown with the DJI Mini 4 Pro drone. Wind speed and direction data were gathered from the drones' built-in IMU sensors during flights. Additional weather data was acquired from the nearby Automated Weather Station (AWS), which functioned as the reference data for comparison (Figs. 6 and 7).

Fig. 4 Arctic Drone Labs (ADL) personnel preparing for drone flights

Fig. 5 Testing drone flights at Hailuoto

Wind speed measurements from the DJI Mavic 3 closely align with AWS data, confirming the drone's capability to capture accurate atmospheric conditions during both flights (see Fig. 6).

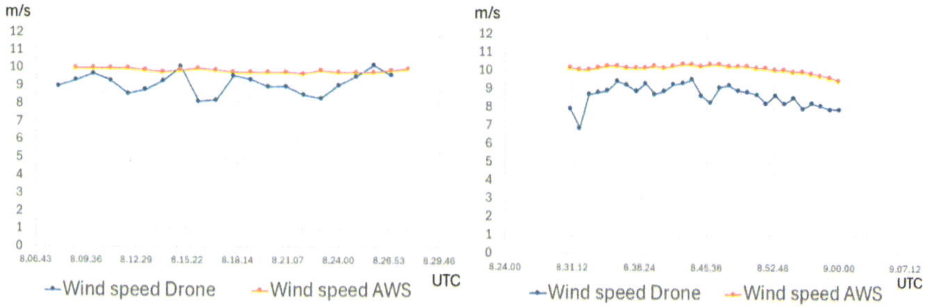

Fig. 6 DJI Mavic 3 wind speed comparison to AWS data from two drone flights

Wind speed measurements from the DJI Mini 4 Pro show strong agreement with AWS data, demonstrating consistent performance across both drone flights (see Fig. 7).

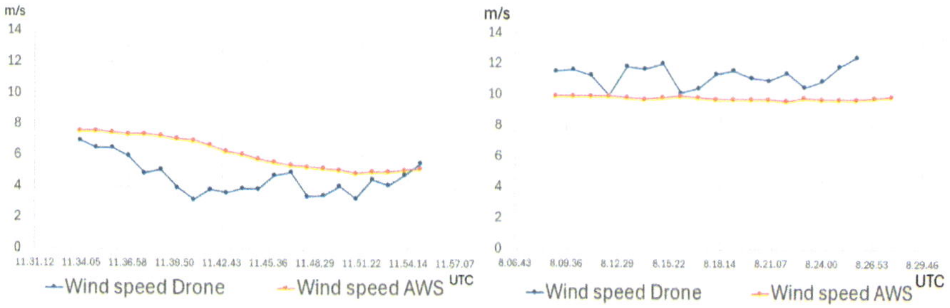

Fig. 7 DJI Mini 4 Pro wind speed comparison to AWS data from two drone flights

For this research, data were gathered during 12 test flights, of which six were flown using a DJI Mavic 3 drone and six were flown using a DJI Mini 4 Pro drone. Wind speed and direction data were gathered via built-in Inertia Measuring Unit (IMU) sensors during flights. As a reference point, additional data from the nearby Automated Weather Station (AWS). These two different data sources were compared from the duration of the flights (Figs. 6 and 7). This research aims to evaluate the precision and quality of weather-related data gathered by drones during flight and to validate these findings by comparing data from established weather stations. Correlation between the data sets displayed in Figs. 6 and 7 was calculated to be the following: during the first flight correlation was 0.35 (Fig. 6 left graph), during the second flight the correlation was 0.60 (Fig. 6 right graph), during the third flight the correlation was 0.53 (Fig. 7 left graph), and during the fourth flight the correlation was 0.67 (Fig. 7 right graph). All results were statistically significant in the statistical correlation analysis (p-value < 0.005). Based on these results, it can be stated that data gathered via drones during flight of the surrounding wind speed are well correlated with similar data collected by stationary weather stations. However, different flights had clear variances in calculated correlation (range 0.35–0.67). This variance can be partly explained by the other properties of the two drones; the small and light drones are more prone to turbulence and changes in wind speed than larger and heavier drones.

The data gathering method depends on the drone's alignment and its state of momentary acceleration related to its location, which enhances the effects of wind gustiness. The elevation level of measurement is a significant factor affecting the gustiness of prevalent wind conditions, which is another factor contributing to the variability of correlations (drone flight elevation during measurement was 29.7 m, and the height of the weather station was 37 m).

During the analysis phase of the research, a new target for future research emerged. A clear gap in knowledge was recognized about low-altitude turbulence and its unpredictable variability. Especially in low altitudes and in proximity to solid structures such as buildings, thick vegetation, or significant variation in the shape of the land surface, can cause significant turbulence, which manifests in gusty wind conditions and irregular airflow surrounding the drone. These phenomena create substantial challenges to flight stability and the opportunities to operate drones at low altitudes or near buildings. The overall characteristics of such weather phenomena are hyper-localized, and mapping or preparing for/reacting to them would require accurate and localized weather measurement.

4 Conclusions

In this research, we explored the possibilities of Unmanned Aircraft Systems (UAS) and especially the use of Inertial Measurement Units (IMUs) in measuring weather conditions. We noticed limited research on enhancing drone operations during wind weather during the literature review, comparing inertial measurement units and localized weather data. The smart and sustainable aviation research group conducted test flights during which data about ongoing wind conditions were measured. Research findings showed that using IMU in measuring weather conditions is promising, but further research is needed to improve its accuracy and reliability. Data from IMU compared to the data from Automated Weather Stations (AWS) was not uniformly correlated. Correlation was observed, but with significant variance. This shows the need for further field testing and improvement.

This research shows that IMU technology can provide a promising possibility for observing and measuring local weather conditions. However, its use is limited, especially in variable weather conditions, requiring further development. The results indicate that drone-based weather observations could be utilized to complement the currently used weather stations and improve weather measurement in areas where conventional weather stations are not suitable.

In future research, more field testing should be conducted in varying and turbulent weather conditions, harsh environments, and different seasons. Additionally, the different properties of different drones and the effects of these differences on weather data gathering should be investigated.

References

1. Abichandani, P., Lobo, D., Ford, G., Bucci, D., & Kam, M. (2020). Wind measurement and simulation techniques in multi-rotor small unmanned aerial vehicles. *IEEE Access, 8*, 54910–54927.

2. Contreras, J. F., Vassaram, J. A., Fernandes, M. R., & do Val, J. B. (2025). Real-time and post-mission heading alignment for drone navigation based on single-antenna GNSS and MEMs-IMU sensors. *Drones, 9*(3), 169.
3. Duffy, J. P., Cunliffe, A. M., DeBell, L., Sandbrook, C., Wich, S. A., Shutler, J. D., Myers-Smith, I. H., Varela, M. R., & Anderson, K. (2018). Location, location, location: Considerations when using lightweight drones in challenging environments. *Remote Sensing in Ecology and Conservation, 4*(1), 7–19.
4. European Union Aviation Safety Agency (EASA). (2023). Drones (UAS). Available at: https://www.easa.europa.eu/en/the-agency/faqs/drones-uas#category-regulations-on-uas-drone-explained. Accessed December 19, 2024.
5. Fagiano, L., Quack, M., Bauer, F., Carnel, L., & Oland, E. (2022). Autonomous airborne wind energy systems: accomplishments and challenges. *Annual Review of Control, Robotics, and Autonomous Systems, 5*(1), 603–631.
6. Finnish Meteorological Institute. (2023). *Setting up a private weather station.* Available at: Private weather station - Finnish Meteorological Institute. Accessed January 1, 2025.
7. Gao, M., Hugenholtz, C. H., Fox, T. A., Kucharczyk, M., Barchyn, T. E., & Nesbit, P. R. (2021). Weather constraints on global drone flyability. *Scientific Reports, 11*(1), 12092.
8. Giersch, S., El Guernaoui, O., Raasch, S., Sauer, M., & Palomar, M. (2022). Atmospheric flow simulation strategies to assess turbulent wind conditions for safe drone operations in urban environments. *Journal of Wind Engineering and Industrial Aerodynamics, 229*, 105136.
9. Julzarika, A., & Harintaka. (2019). Free global DEM: Converting DSM to DTM and ITS applications. *The International Archives of the Photogrammetry, Remote Sensing and Spatial Information Sciences, 42*, 319–325.
10. Laghari, A. A., Jumani, A. K., Laghari, R. A., & Nawaz, H. (2023). Unmanned aerial vehicles: A review. *Cognitive Robotics, 3*, 8–22.
11. Meier, K., Hann, R., Skaloud, J., & Garreau, A. (2022). Wind estimation with multirotor UAVs. *Atmosphere, 13*(4), 551.
12. Panttila, H., & Valta R.-O. (2022). *Droonit tutkimuskäytössä.* Available at: Droonit tutkimuskäytössä | GTK. Accessed January 8, 2025.
13. Roseman, C. A., Argrow, B. M., & Pinto, J. O. (2019). Targeted weather forecasts for small unmanned aircraft systems. In *19th Conference on Aviation, Range, and Aerospace Meteorology, Phoenix, AZ*, Amer. Meteor. Soc (Vol. 1).
14. SESAR Joint Undertaking. (2016). *European Drones Outlook Study.* Available at: https://www.sesarju.eu/sites/default/files/documents/reports/European_Drones_Outlook_Study_2016.pdf. Accessed January 7, 2025.
15. Traficom. (2023). *The drone and its functions – UAS general knowledge.* Available at: https://droneinfo.fi/en/study-material/drone-and-its-functions-uas-general-knowledge. Accessed December 19, 2024.
16. Tuominen, S., Koivula, H., Kahiluoto, J., Haapanen, K., Pirinen, M., & Hirvonen, J. (2021). Kulttuuriympäristöjen ja maisemien visuaalisen seurannan menetelmät. *Ympäristöministeriö, Helsinki. Ympäristöministeriön julkaisuja, 36.*

Swarm Intelligence and Autonomous Coordination

Persistent Monitoring of Large Dispersed Areas Using Drone Swarms: Algorithms and Experimental Validation

Muhamad Rausyan Fikri[1(✉)], Eetu Silvennoinen[1], Made Widhi Surya Atman[2], Yury Nikulin[3], and Azwirman Gusrialdi[1]

[1] Automation Technology and Mechanical Engineering, Faculty of Engineering and Natural Sciences, Tampere University, Tampere, Finland
muhamad.fikri@tuni.fi , azwirman.gusrialdi@tuni.fi
[2] Robotics and Autonomous Systems, University of Turku, Turku, Finland
[3] Department of Mathematics and Statistics, University of Turku, Turku, Finland

Abstract. This paper presents a two-layer control architecture for coordinating a swarm of drones equipped with heterogeneous sensing capabilities to efficiently and continuously monitor targeted areas distributed across a large area. In the first layer, the targeted areas are clustered into several regions approximated as convex polygons, where the drones will be allocated to these regions. For this allocation, a novel optimization problem is formulated to balance the trade-off between minimizing uncovered areas and ensuring equitable coverage across regions. The second layer enables drones to persistently and safely monitor the targeted areas within their assigned regions. Simulation and experimental results validate the effectiveness of the proposed architecture using a small-scale drone swarm to demonstrate the monitoring performance, including enhanced coverage and reduced travel distances for the drones.

Keywords: Multiobjective optimization problem · Drone swarms task allocation · Persistent monitoring

1 Introduction

Persistent monitoring involves the continuous inspection of specific, targeted areas across large, dispersed areas, such as crop health monitoring in agriculture and environmental monitoring in forestry [1,2]. Swarms of multirotor drones, equipped with heterogeneous sensors, are the key enablers for automating this task. In contrast to single drones, they can cover vast areas more efficiently and provide robustness to system failures [3]. In practice, drones have limited battery capacity [4], and their total sensing area is significantly smaller than the total size of the area to be covered. This calls for a cooperative control algorithm, enabled via wireless communication technologies, to coordinate the drone swarms and ensure effective coverage of the large, dispersed areas while ensuring that no collisions occur among the drones.

M. Laakkonen (ed.), *DroneAI for Resilient Nature*,
https://doi.org/10.1007/978-3-032-19008-6_5

Numerous cooperative control algorithms for the persistent monitoring of large areas using drone swarms have been proposed in the literature (e.g., [5–8]). However, in practical scenarios, the distribution of targeted areas is often uneven across the monitored area. Consequently, poorly designed control strategies may lead to drones becoming trapped in local optima, as demonstrated in [9], resulting in inefficient energy usage and unnecessary monitoring of less critical areas instead of focusing on the targeted areas. Unfortunately, this issue has been largely overlooked in the existing work.

In this paper, we adopt and extend the two-layer control architecture (TLCA) introduced in our previous work [10] to address the aforementioned gap. The first layer focuses on minimizing battery usage, specifically by reducing the travel distances of drones. This is achieved by clustering the targeted areas into several regions approximated as convex polygons and assigning drones to these regions. In contrast to [10,11], we propose a novel optimization problem that balances the trade-off between minimizing uncovered areas and ensuring equitable coverage across regions. Additionally, we introduce a convex optimization using mixed-integer linear programming (MILP) as an alternative method. The second layer enables drones to continuously and safely monitor all targeted areas within their assigned regions, avoiding collisions. The proposed TLCA is validated through Monte Carlo simulations, which explicitly examine the trade-off between uncovered areas and equitable coverage, as well as assess drone travel distances and computational time under varying sensing areas and numbers of regions. Finally, the TLCA is implemented using small-scale drones, demonstrating its practical applicability to real-world persistent monitoring tasks.

2 Problem Formulation

This section explains the problem setting and the necessary notations for developing TLCA. We use $\mathbf{I}_n$ to denote $n \times n$ identity matrix, and 1_n denotes an all-one column vector of size n. All matrix notations are written in bold uppercase letters and vectors with lowercase letters. Moreover, we denote $\mathbb{R}^n$ and $\mathbb{Z}^n$ as real and integer vectors of size n, respectively. Let $\lambda_{\min}(\mathbf{A})$ and $\lambda_{\max}(\mathbf{A})$ denote the eigenvalues of the matrix $\mathbf{A}$ with the smallest and the biggest real part, $\otimes$ denotes the Kronecker product, and the superscript T denotes a transpose.

We consider a scenario of persistent monitoring using drone swarms in a large area, given by a convex polytope-shaped region (compact set) $\mathcal{F} \subset \mathbb{R}^2$, as illustrated in Fig. 1 (left). The field $\mathcal{F}$ is discretized into square grid cells with equal width D_s where some grids of interest need to be monitored (also referred to as *targeted areas*). The persistent monitoring is carried out by n drones, indexed by $\mathcal{I} = \{1, \ldots, n\}$. We assume that the swarms are equipped with heterogeneous sensors that measure the same modality of information. Specifically, each drone i is assumed to sense a subset of the area in the field within a fixed sensing radius $R_i > 0$, i.e., a sensing area of $s_i := \pi R_i^2 \in \mathbb{R}_{>0}$. For example, this scenario encompasses swarms equipped with downward-facing cameras (or LiDARs) with different sensor resolutions. For the case of a downward-facing camera, each drone's sensing radius scales proportionally to its flight altitude and its image plane, and inversely proportional to its focal length, see [14]. Here, without loss of generality, we assume that each robot's sensing radius encapsulates the optimal sensor configuration for data acquisition. Hence, each drone maintains its corresponding state, ensuring a consistent sensing radius; for example, a camera-based drone maintains its altitude throughout the mission. Our goal in this study is to develop a control architecture that includes a control algorithm to coordinate n drones to continuously monitor and maximize coverage of targeted areas while minimizing the travel distance of the drones, thereby reducing energy consumption.

3 Two-Layer Control Architecture

To achieve the goal, we adopt and adapt a two-layer control architecture (TLCA), initially proposed in [10], which consists of task allocation and persistent monitoring strategies for drone swarms, as illustrated in Fig. 1 (Right). In the first layer, K-means clustering divides the targeted areas within the field $\mathcal{F}$ into ℓ (pre-defined by the user) clusters. A convex hull then approximates the boundary of each cluster, see [10] for more details. The region (convex hull) associated with the jth cluster is denoted by $\mathcal{Q}^j$. Moreover, we consider the case $\ell \leq n$. The drones are then assigned to these regions as described in the following subsection.

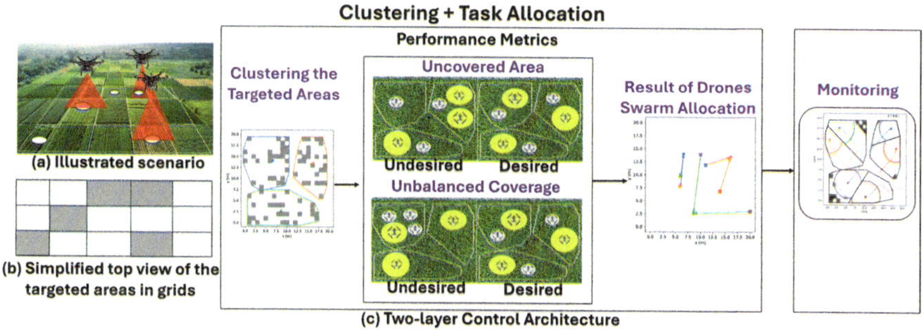

Fig. 1. (Left) The scenario of a field monitored by drone swarms equipped with sensors. The white dot object represents the targeted point, and the gray color indicates a simplified view of the targeted areas. (Right) Two-layer control architecture for persistent monitoring of drone swarms

3.1 Drone Swarms Task Allocation

The objective is to allocate each drone to one of the regions $\mathcal{Q}^j$ to minimize uncovered areas while balancing coverage between areas. This problem can be formulated as the following optimization problem:

$$\underset{x \in \mathcal{X}}{\text{minimize}} \quad \beta_f f(x) + \beta_g g(x). \tag{1}$$

Remark 1. *While the individual metrics $f(x)$ and $g(x)$ were optimized separately in [11], here we merge them into a single multi-objective formulation to explicitly explore the trade-off between total uncovered area and coverage balance.*

Here, $x = (x_{11}, \ldots, x_{1\ell}, \ldots, x_{n1}, \ldots, x_{n\ell})^T \in \{0, 1\}$ and the scalar values $\beta_f, \beta_g \geq 0$, pre-defined by the user, dictates the trade-off between the two cost functions. The value $x_{ij} = 1$ denotes a drone i to region j and $x_{ij} = 0$ otherwise. Moreover, the set constraints $\mathcal{X}$ is defined as

$$\mathcal{X} := \left\{ x_{ij} \in \{0,1\} \middle| \sum_{j=1}^{\ell} x_{ij} = 1 \, \forall i = 1, \ldots, n, \sum_{i=1}^{n} x_{ij} \geq 1 \;\, \forall j = 1, \ldots, \ell \right\}. \tag{2}$$

The equality constraint in (2) ensures that each drone is assigned to a region, while the inequality guarantees that at least one drone monitors a region. The cost function $f(x)$ penalizes regions with both over-coverage and under-coverage and is defined as [10]:

$$f(x) = \sum_{j=1}^{\ell} \left(c_j - \sum_{i=1}^{n} s_i x_{ij} \right)^2, \tag{3}$$

where, c_j is the area of the region. On the other hand, cost function $g(x)$ measures the unbalanced coverage and is given by [11,12].

$$g(x) = \sum_{(j,k),\, j \neq k} \left(\sum_{i=1}^{n} \frac{s_i}{c_k} x_{ik} - \sum_{i=1}^{n} \frac{s_i}{c_j} x_{ij} \right)^2, \; j, k = 1, \ldots, \ell. \tag{4}$$

The cost function $f(x)$ can be written in a compact form as:

$$f(x) = \frac{1}{2} x^T \mathbf{Q}_f x + p_f^T x + r_f, \tag{5}$$

where $\mathbf{Q}_f = 2(\mathbf{I}_\ell \otimes s s^T)$, $p_f^T = -2c^T(\mathrm{I}_\ell \otimes s^T)$, and $r_f = c^T c$. Similarly, the cost function $g(x)$ can also be written as

$$g(x) = \frac{1}{2} x^T \mathbf{Q}_g x + p_g^T x + r_g, \tag{6}$$

where $\mathbf{Q}_g = 2(\mathbf{C} \otimes s)\mathbf{D}^T\mathbf{D}(\mathbf{C} \otimes s^T)$, $p_g^T = 0_n$, and $r_g = 0$. Here, we denote $\mathbf{C} = \mathrm{diag}(\frac{1}{c_1}, \ldots, \frac{1}{c_\ell})$, and $\mathbf{D}^T\mathbf{D} = 2(\ell \mathbf{I}_\ell - 1_\ell 1_\ell^T)$ denotes a matrix of j and k pair whereas $j \neq k$. Note that matrices $\mathbf{Q}_f$ and $\mathbf{Q}_g$ are symmetric and indefinite. Finally, we can write the constraint set (2) as

$$\mathcal{X} := \{x \in \{0,1\}^{n\ell} \mid \mathbf{A}_1 x = b_1, \;\; \mathbf{A}_2 x \geq b_2\}. \tag{7}$$

where $\mathbf{A}_1 = \mathbf{I}_n \otimes 1_\ell^T$, $\mathbf{A}_2 = 1_n^T \otimes \mathbf{I}_\ell$, $b_1 = 1_n$ and $b_2 = 1_\ell$. The following subsections present two methods for solving (1) by transforming optimization (1) into concave quadratic programming and mixed-integer linear programming.

Concave Quadratic Formulation: Let us consider the following optimization problem:

$$\underset{x \in \mathcal{X}}{\text{minimize}} \quad \beta_f \bar{f}(x) + \beta_g \bar{g}(x), \tag{8}$$

where $\bar{f}(x) = \frac{1}{2}x^T\bar{\mathbf{Q}}_f x + p_f x + r_f$ and $\bar{g}(x) = \frac{1}{2}x^T\bar{\mathbf{Q}}_g x + p_g x + r_g$. Matrix $\bar{\mathbf{Q}}_f$ and $\bar{\mathbf{Q}}_g$ are defined as $\bar{\mathbf{Q}}_f = \mathbf{Q}_f - \alpha_f \mathbf{I}$ and $\bar{\mathbf{Q}}_g = \mathbf{Q}_g - \alpha_g \mathbf{I}$, where $\alpha_f = \lambda_{\max}(\mathbf{Q}_f) + \epsilon$, $\alpha_g = \lambda_{\max}(\mathbf{Q}_g) + \epsilon$ and $\epsilon > 0$ any small shift constant.

Note that the cost functions $\bar{f}(x)$, $\bar{g}(x)$ are concave and as a result $\beta_f \bar{f}(x) + \beta_g \bar{g}(x)$ is also a concave function. Furthermore, we have $\beta_f \bar{f}(x) + \beta_g \bar{g}(x) = (\beta_f f(x) + \beta_g g(x)) - 0.5(\beta_f \alpha_f + \beta_g \alpha_g)\|x\|_2^2$. Due to the equality constraint in $\mathcal{X}$ which ensures all drones are assigned to one of the regions, the terms $0.5(\beta_f \alpha_f + \beta_g \alpha_g)\|x\|_2^2$ is constant for a given α_f, α_g, and β_f, $\beta_g \geq 0$ and for any feasible candidate solution. Hence, the optimal solution to (8) is the same as the optimal solution to (1). Moreover, note that the constraint (7) is totally unimodular [11]. As a result, the optimization solution (8) can be obtained by relaxing the binary variable $\mathcal{X}$ into $\mathcal{D} = \{x \in [0,1]^{n\ell} \mid \mathbf{A}_1 x = b_1, \ \mathbf{A}_2 x \geq b_2\}$.

Convex Mixed-Integer Linear Programming: Let us consider the following convex Mixed-Integer Linear Programming (MILP) problem:

$$\underset{x \in \mathcal{X}}{\text{minimize}} \quad \beta_f \tilde{f}(x) + \beta_g \tilde{g}(x), \tag{9}$$

where $\tilde{f}(x) = \frac{1}{2}x^T\tilde{\mathbf{Q}}_f x + p_f x + r_f$ and $\tilde{g}(x) = \frac{1}{2}x^T\tilde{\mathbf{Q}}_g x + p_g x + r_g$. Matrices $\tilde{\mathbf{Q}}_f$ and $\tilde{\mathbf{Q}}_g$ are defined as $\tilde{\mathbf{Q}}_f = \mathbf{Q}_f - \alpha_f \mathbf{I}$ and $\tilde{\mathbf{Q}}_g = \mathbf{Q}_g - \alpha_g \mathbf{I}$, where $\alpha_f = \lambda_{\min}(\mathbf{Q}_f) - \epsilon$, $\alpha_g = \lambda_{\min}(\mathbf{Q}_g) - \epsilon$ and $\epsilon > 0$ any small shift constant.

Note that cost functions $\tilde{f}(x)$, $\tilde{g}(x)$ are convex, and as a result $\beta_f \tilde{f}(x) + \beta_g \tilde{g}(x)$ is also a convex function. Using the same argument for (8), the optimal solution to (9) is the same as the optimal solution to (1).

While the constraints in (7) are totally unimodular, it does not guarantee that the solution to the relaxed convex optimization problem (i.e., relaxing the constraint $\mathcal{X}$ to $\mathcal{D}$) is binary. Therefore, we apply a well-known reformulation-linearization technique (RLT) to find the optimal solution to (9). To this end, we introduce a new binary variable y is a vector which collects all y_{ijkl} to replace the bilinear terms x_{ij} by $y_{ijkl} = x_{ij}x_{kl}$, and reformulate the constraints as follow:

$$\mathcal{Z} = \left\{ (x,y) \,\middle|\, \begin{array}{llll} A_1 x = b_1, & A_2 x \geq b_2, & y_{ijkl} \geq x_{ij} + x_{kl} - 1, & \\ y_{ijkl} \geq 0, & y_{ijkl} \leq x_{ij}, & y_{ijkl} \leq x_{kl}, & x_{ij} \in \{0,1\} \end{array} \right\}. \tag{10}$$

The cost functions can then be written as:

$$\tilde{f}(x,y) = \sum_{i=1}^{n}\sum_{j=1}^{\ell}\sum_{k=1}^{n}\sum_{l=1}^{\ell} \tilde{\mathbf{Q}}_{f_{ij}} y_{ijkl} + \sum_{i=1}^{n}\sum_{j=1}^{\ell}\left(\frac{1}{2}\tilde{\mathbf{Q}}_{f_{ii}} + p_{ij}\right) x_{ij} + r, \tag{11a}$$

$$\tilde{g}(x,y) = \sum_{i=1}^{n}\sum_{j=1}^{\ell}\sum_{k=1}^{n}\sum_{l=1}^{\ell} \tilde{\mathbf{Q}}_{f_{ij}} y_{ijkl} + \sum_{i=1}^{n}\sum_{j=1}^{\ell}\left(\frac{1}{2}\tilde{\mathbf{Q}}_{g_{ii}} + p_{ij}\right) x_{ij} + r, \tag{11b}$$

Finally, the optimization problem (9) can be reformulated as the following mixed-integer linear programming:

$$\text{minimize } \beta_f \tilde{f}(x,y) + \beta_g \tilde{g}(x,y) \tag{12}$$

$$\text{Subject to: } \quad (x,y) \in \mathcal{Z}. \tag{13}$$

3.2 Persistent Monitoring Algorithm

Here, we revisit the persistent monitoring algorithm for the lower layer that dictates the drones' control input to monitor the targeted areas cooperatively. The algorithm follows closely from [10] and is summarized here for each region $\mathcal{Q}^j$, $\forall j \in \mathcal{C}$. Additionally, a proportional gain ψ is introduced here as the scaling gain parameter, allowing adjustment of the resulting control input to match the drone's actuator range. First, the importance of all locations within $\mathcal{Q}^j$ is modeled as a time-varying density function $\phi^j : \mathcal{Q}^j \to [0,\ 1]$ and initialized via a mixture of Gaussian functions ([10], Eq. 5), where 1 specifies the location with the highest importance to monitor, i.e., the targeted areas, and 0 denotes the location with the least importance. The density function ϕ^j over $\mathcal{Q}^j$ is continuously updated according to the drones' location. To be more precise, the density value of a point $q \in \mathcal{Q}^j$ decreases with a fixed rate $\underline{\delta}$ when q is being monitored by at least one drone, and increases with the rate $\bar{\delta}$ if q is left unmonitored, requiring the drones to revisit that point to maintain persistent monitoring periodically. The parameters $\underline{\delta}$ and $\bar{\delta}$ are adjusted according to the characteristics of the sensor attached to the drones.

Let us define the set of selected drones to monitor $\mathcal{Q}^j$ as $\mathcal{I}^j$. Moreover, the Voronoi partition $\mathcal{V}_i(p_i) = \{q \in \mathcal{Q}^j \mid \|q - p_i\| \leq \|q - p_k\| \ \forall k \in \mathcal{I}^j \setminus \{i\}\}$ dynamically assigns each point in $\mathcal{Q}^j$ to drone i. We denote the *sensing region* of drone i at position p_i by $\mathcal{C}_i(p_i) = \left\{q \in \mathbb{R}^2 \mid \|q - p_i\| \leq R_i\right\}$, where R_i is the maximum sensing radius of drone i. Finally, we define the *feasible sensing region of drone i* $\mathcal{S}_i(p_i)$ as the portion of $\mathcal{C}_i(p_i)$ that lies within the Voronoi cell, i.e., $\mathcal{S}_i(p_i) = \mathcal{C}_i(p_i) \cap \mathcal{V}_i(p_i)$, where each drone i can and needs to gather data via its sensor.

Furthermore, it is assumed that the kinematic model of each drone is given by:

$$\dot{p}_i = u_i \tag{14}$$

where $u_i \subseteq \mathbb{R}^2$ denotes the velocity input to be designed. While simplistic, the model in (14) describes multirotor drones with a low-level controller that tracks the velocity input u_i. The control input u_i for all drone $i \in \mathcal{I}^j$ is designed as

$$
\begin{aligned}
u_i &= \begin{cases} 2\psi \text{mass}(\mathcal{S}_i(p_i))(\text{cent}(\mathcal{S}_i(p_i)) - p_i), & \text{if } p_i \in \mathcal{Q}^j \\ \gamma(c^j - p_i) & \text{otherwise,} \end{cases} \\
\text{mass}(\mathcal{S}_i) &= \int_{\mathcal{S}_i} \phi^j(q,t)\,\mathrm{d}q, \qquad \text{cent}(\mathcal{S}_i) = \frac{1}{\text{mass}(\mathcal{S}_i)} \int_{\mathcal{S}_i} q\phi^j(q,t)\,\mathrm{d}q,
\end{aligned} \tag{15}
$$

where $\gamma > 0$ is the proportional gain. The control input u_i is formulated as a combination of gradient ascent control to maximize the coverage over the feasible sensing area $\mathcal{S}_i(p_i)$ and a proportional control towards the centroid of $\mathcal{Q}^j$, i.e., c_j, when the drone is outside $\mathcal{Q}^j$. Each drone distributes its control input using local information, while the density function $\phi^j(q,t)$ is updated centrally. Finally, as presented in [10], distances between the drones can be maintained above a predefined safety distance (e.g., to avoid collisions with each other or to reduce downwash effect), and actuator limits can be easily adopted using a quadratic programming-based approach via control barrier functions.

4 Numerical Simulation and Discussion

In this section, we evaluate the TLCA via numerical simulations, focusing on four key performance assessments: task-allocation computational time (comparing concave quadratic versus MILP formulations); coverage performance (ratio of uncovered areas over time); total drone travel distance; and the objective trade-off between minimizing $f(x)$ and balancing $g(x)$.

4.1 Simulation Setup

Consider a field $\mathcal{F}$ of size $[-0.5, 19.5]\,\text{m} \times [-0.5, 19.5]\,\text{m}$ discretized into 400 grids with grid size $D_s = 1.0\,\text{m}^2$ and 20% of grids are designated as the targeted areas. Additionally, the subinterval size for discretizing the integration domain in (15) is set to 0.2. The performance of drone swarms' task allocation algorithms in TLCA is evaluated under two scenarios: (1) uniform distribution targeted areas throughout $\mathcal{F}$ denoted by $\mathcal{H}^{\text{u}}$, and (2) clustered distribution targeted areas towards the four corners of the field $\mathcal{F}$ denoted by $\mathcal{H}^{\text{c}}$, see Fig. 2. We deploy eight drones with different sensing areas. Table 1 shows the number of regions with $\ell = 1, \ldots, 4$. Finally, we perform a Monte Carlo simulation by generating 50 sets of random sensing areas s_i whose values are varied from 4–13 m^2 for task allocation and 20 sets for coverage performance evaluation.

Table 1. Sizes of partitioned regions (in m^2) from K-means clustering

$\mathcal{H}^{\text{u}}$ (Uniform distribution)						$\mathcal{H}^{\text{c}}$ (Clustered distribution)				
	1	2	3	4	Total	1	2	3	4	Total
$\ell = 1$	377.0				377.0	372.0				372.0
$\ell = 2$	187.0	165.5			352.5	151.0	110.5			261.5
$\ell = 3$	117.5	87.5	119.0		324.0	35.0	110.5	46.0		191.5
$\ell = 4$	62.0	97.5	95.5	59.0	314.0	35.0	37.0	46.0	34.5	152.5

We evaluate drone swarms task allocation formulated using a concave quadratic (8) and an MILP (12) with three weights values settings: (1) $\beta_f = 0.0$, $\beta_g = 1.0$, (2) $\beta_f = 1.0$, $\beta_g = 0.0$, and (3) $\beta_f = 0.2$, $\beta_g = 1000$. The evaluation focuses on comparing the computational efficiency of allocating drones to designated areas (runtime and variance), coverage performance based on the ratio of uncovered targeted areas, and drones' travel distance to assess the trade-offs between minimizing uncovered areas $f(x)$ and balancing coverage $g(x)$. First, we introduce the following definition to verify whether the solutions to $f(x)$ and $g(x)$ are conflicting.

Definition 1. *Let* $x_f^* = \arg\min_{x \in X} f(x)$ *and* $x_g^* = \arg\min_{x \in X} g(x)$. *The two objectives* $f(x)$ *and* $g(x)$ *are* in conflict *if and only if* $f(x_f^*) \neq f(x_g^*)$ *and* $g(x_f^*) \neq g(x_g^*)$.

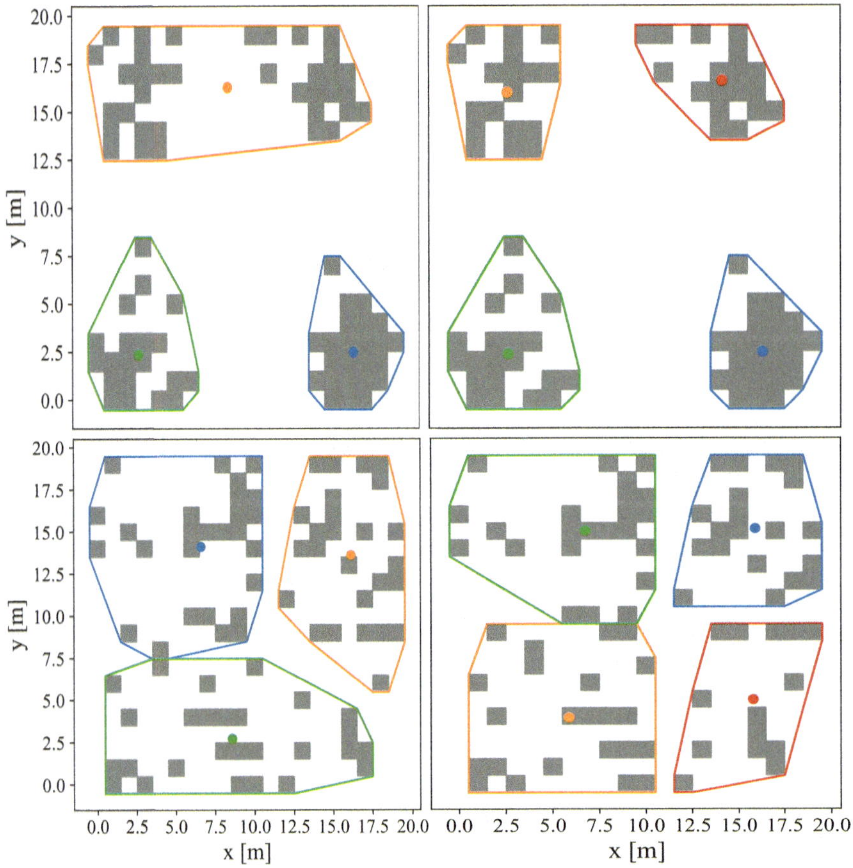

Fig. 2. The illustration of clustering using K-means for targeted areas that are clustered (top) and uniformly distributed (bottom), each with 3 and 4 clusters

4.2 Evaluation of Drone Swarms Task Allocation Strategy

In this section, we compare the computational time for solving both concave quadratic (8) and MILP (12) under uniform ($\mathcal{H}^{\mathrm{u}}$) and clustered ($\mathcal{H}^{\mathrm{c}}$) distributions. Table 2 presents the mean runtime ± variance over 50 Monte Carlo runs for three weight settings—$(\beta_f, \beta_g) = (1.0, 0.0), (0.0, 1.0), (0.2, 1000)$—and the number of regions $\ell = 2, 3, 4$. The results in Table 2 show that the concave quadratic programming consistently outperforms MILP; for example, when $(\beta_f, \beta_g) = (1.0, 0.0)$, its mean runtime on uniform distribution $\mathcal{H}^{\mathrm{u}}$ ranges from 0.05 ± 0.01 to $0.15\,\mathrm{s} \pm 0.03$, compared to MILP's $0.70\,\mathrm{s} \pm 0.59$ to $2.20\,\mathrm{s} \pm 0.19$. On clustered distribution $\mathcal{H}^{\mathrm{c}}$, the concave quadratic programming runs in $0.02\,\mathrm{s} \pm 0.01$ to $0.06\,\mathrm{s} \pm 0.02$, compared to MILP's $0.71\,\mathrm{s} \pm 0.48$ to $5.20\,\mathrm{s} \pm 1.16$. We also verify the conflict between our two objectives via Definition 1. Specifically, let x_f^* and x_g^* be the minimizes of $f(x)$ and $g(x)$, respectively. On all 50 Monte Carlo sets of random sensing areas, we observe $f(x_f^*) \neq f(x_g^*)$ and $g(x_f^*) \neq g(x_g^*)$ in every case, confirming that neither objective dominates the other and thus that $f(x)$ and $g(x)$ are indeed in conflict.

Table 2. The average and variance of runtime for solving task allocation (s)

			$\mathcal{H}^{\mathrm{u}}$ (Uniform)			$\mathcal{H}^{\mathrm{c}}$ (Clustered)		
β_f	β_g	Metric	2	3	4	2	3	4
0.0	1.0	Average (8)	0.16	2.26	6.43	0.19	2.32	6.73
		Variance (8)	0.04	0.72	2.24	0.09	0.62	2.18
		Average (12)	1.51	2.33	15.60	1.50	2.42	15.24
		Variance (12)	0.71	0.35	2.65	0.69	0.43	2.61
0.2	1000	Average (8)	0.07	1.17	3.77	0.08	1.41	4.19
		Variance (8)	0.01	0.15	0.56	0.02	0.32	0.76
		Average (12)	1.09	2.27	14.14	1.38	2.35	15.40
		Variance (12)	0.86	0.29	2.27	1.30	0.43	2.56
1.0	0.0	Average (8)	0.05	0.12	0.15	0.02	0.03	0.06
		Variance (8)	0.01	0.04	0.03	0.01	0.01	0.02
		Average (12)	0.70	1.24	2.20	0.71	0.93	5.20
		Variance (12)	0.59	0.10	0.19	0.48	0.15	1.16

4.3 Evaluation on Persistent Monitoring Algorithm

Next, we analyze coverage performance under two different scenarios defined in Sect. 4.1, uniform ($\mathcal{H}^{\mathrm{u}}$) and clustered ($\mathcal{H}^{\mathrm{c}}$) distribution, for number of regions $\ell = 3$ and 4, with three weight settings $(\beta_f, \beta_g) = (1.0, 0.0)$, $(0.0, 1.0)$, $(0.2, 1000)$. We quantify coverage at time t by the ratio of uncovered targeted areas, computed as the number of targeted areas not yet visited divided by the total number of targeted grid areas, which ranges from 1 (no area covered) down to 0 (all areas covered). Figure 3 shows the coverage performance with $\ell = 3$ and $\ell = 4$. An additional zoom-in result is attached to $\ell = 3$ in Fig. 3 to further the observation of conflict in the coverage performance result. In the clustered distribution with $\ell = 3$, the coverage performance under ($\beta_f = 0.0,\ \beta_g = 1.0$) is comparable to that under ($\beta_f = 1.0,\ \beta_g = 0.0$). For $\ell = 4$, however, ($\beta_f = 1.0,\ \beta_g = 0.0$) significantly outperforms ($\beta_f = 0.0,\ \beta_g = 1.0$). In the uniform distribution, at $\ell = 3$ the weights ($\beta_f = 0.0,\ \beta_g = 1.0$) outperform ($\beta_f = 1.0,\ \beta_g = 0.0$), whereas at $\ell = 4$ the opposite holds: ($\beta_f = 1.0,\ \beta_g = 0.0$) outperforms ($\beta_f = 0.0,\ \beta_g = 1.0$). The result in Fig. 3 shows that $f(x)$ and $g(x)$ remain in conflict under both distribution scenarios. Thus, neither function dominates the other. Hence, by introducing the multiobjective formulation (1), for example with the weights ($\beta_f = 0.2, \beta_g = 1000$), we obtain a non-dominated coverage performance between $f(x)$ and $g(x)$, where the result is between $\beta_f = 1.0,\ \beta_g = 0.0$ and $\beta_f = 0.0,\ \beta_g = 1.0$ (also illustrated in Fig. 3).

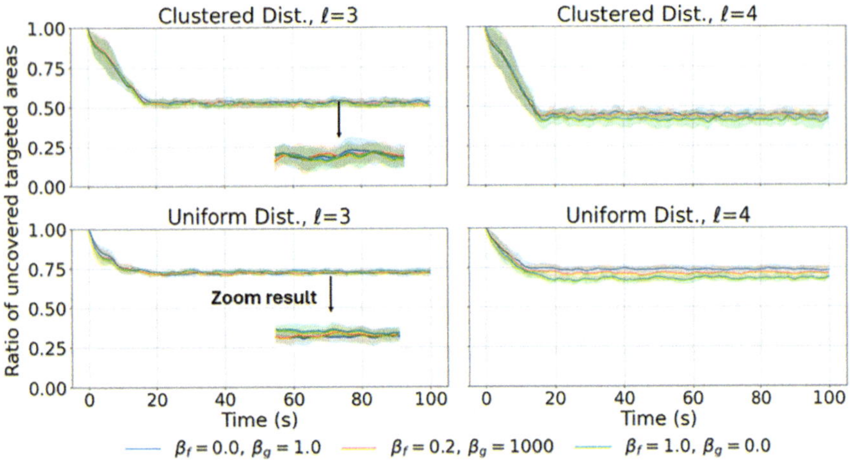

Fig. 3. Comparison of coverage performance under uniform and clustered target-area distributions for three different weights $(\beta_f, \beta_g) = (0.0, 1.0), (1.0, 0.0), (0.2, 1000)$. The results show distinct coverage performance for each weight, demonstrating that a difference in weight leads to different swarm allocations and, consequently, affects coverage performance

Next, we assess how clustering and task allocation (the first TLCA layer) affect the travel distance of drones under both uniform and clustered target-area scenarios (Fig. 4). Clustering and allocation reduce travel for most drones; however, under $\ell = 3, 4$, drones 2 and 3 travel farther than other drones. Recall that coverage is executed with a Voronoi-based coverage control algorithm. In scenario $\ell = 1$, for both uniform and clustered targeted areas, this algorithm

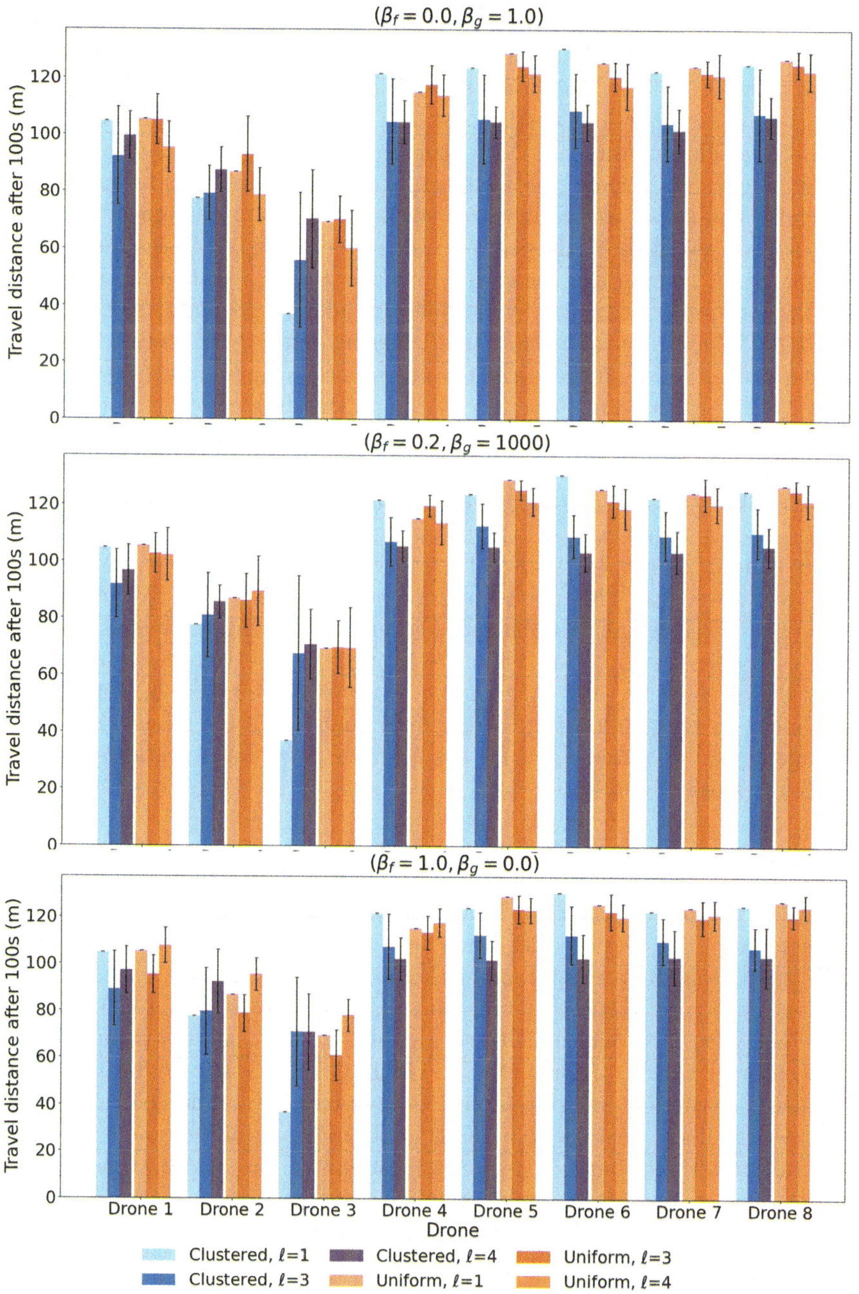

Fig. 4. The drone swarms traveling distance for different distribution scenarios with varying weights $(\beta_f, \beta_g) = (0.0, 1.0), (1.0, 0.0), (0.2, 1000)$ and fixed in $\ell = 3$ and $\ell = 4$

balances the coverage area. Consequently, when drones 2 and 3 have a smaller sensing range, their travel distance is limited, as is their ability to sense the nearest targets. When ℓ increases to 3 or 4, the large region that all drones share at $\ell = 1$ is split into smaller clusters. In these smaller clusters, drones 2 and 3 are no longer dominated by other drones with larger sensing areas. The reduced cluster size brings the next target within their limited sensing range, a plausible condition, especially when drones 2 and 3 are assigned to different clusters. As a result, both drones can visit more targets and accumulate longer flight paths. A comparison between $\ell = 3$ and $\ell = 4$ reveals a clear difference Fig. 3, which can be examined by considering the reduced cluster sizes in Table 1 and Fig. 2. When a drone with a small sensing range is assigned alone to the smallest cluster region, its required travel distance is longer in the $\ell = 3$ scenario because the gap between the region's size and the drone's sensing area is large. In contrast, under $\ell = 4$, the smaller gap between the size of the region and the sensing area of the drone allows drones to cover their assigned regions with a shorter travel distance. The coverage performance, influenced by the weights β_f and β_g, directly contributes to the travel distance of drones. For instance, with $\beta_f = 0.0$ and $\beta_g = 1.0$, drones 2 and 3 travel less in the $\ell = 4$ scenario compared to the other weights. This performance occurs due to $\beta_f = 0.0$ and $\beta_g = 1.0$, which prioritize minimizing unbalanced coverage, allowing the drones' allocation to appear under the condition where the total travel distance of all drones is less than that of other β_f and β_g weights. In contrast, for $\ell = 3$, obtaining the unbalanced coverage allocation differs from that for $\ell = 4$. Thus, the result of coverage in $\beta_f = 0.0$ and $\beta_g = 1.0$ has a longer total travel distance compared to other weights β_f and β_g.

5 Experimental Validation and Discussion

Finally, we conducted experiments and simulations to validate and compare the performance of our proposed TLCA. Four Crazyflies drones ($92 \times 92 \times 29$ mm) are deployed in an indoor environment with dimensions of $2\,\text{m} \times 2\,\text{m}$. Each drone operated at a fixed altitude of 1 m, with sensing areas of $[1.0, 1.1, 1.2, 1.3]\,\text{m}^2$. Drone positions are continuously tracked by six Vicon motion cameras, where the position data is transmitted through the Vicon Bridge to a ROS 2-based centralized controller, integrated with Crazyswarm [13]. The central controller computes the linear velocity commands at 50 Hz based on the model in (14) and then transmits these commands wirelessly (within 20 ms) via two Crazyflie radio dongles, each managing communication with two drones.

For comparative analysis, we created a simulation environment to replicate the experimental conditions from the indoor space dimension, initial drone positions, sensing areas, and some algorithm parameters. However, in simulation, we assume the instantaneous inter-drone communication, the absence of latency in command, and perfect position accuracy. These simplifications provide a baseline for evaluating the real-world performance.

The grid size is reduced from $D_s = 1.0$ to $0.1\, textm^2$, the gradient ascent gain $\psi = 5$, and the subinterval size for discretization of integration domain in

(15) is changed from 0.2 to 0.8. The scenario involved a uniform distribution of 80 target areas with weights $(\beta_f, \beta_g) = (1.0, 0.0)$, where $\ell = 2$. All simulations and experiments utilize a central computer with an Intel Core i7 processor and 32 GB of RAM.

Figure 5a shows the simulation trajectories at time intervals $t = 0$–10 s, 10.1–20 s, and 20.1–30 s. Figure 5b shows the simulation trajectories through RViz at the same time intervals and parameters as in the simulation version, and we reconstruct the experimental trajectory plot into a Python GUI to illustrate the navigation paths of drones over the targeted areas. Figure 5c compares the ratio of uncovered regions over time (0–30 s) in a line plot with additional zooming result next to the figure.

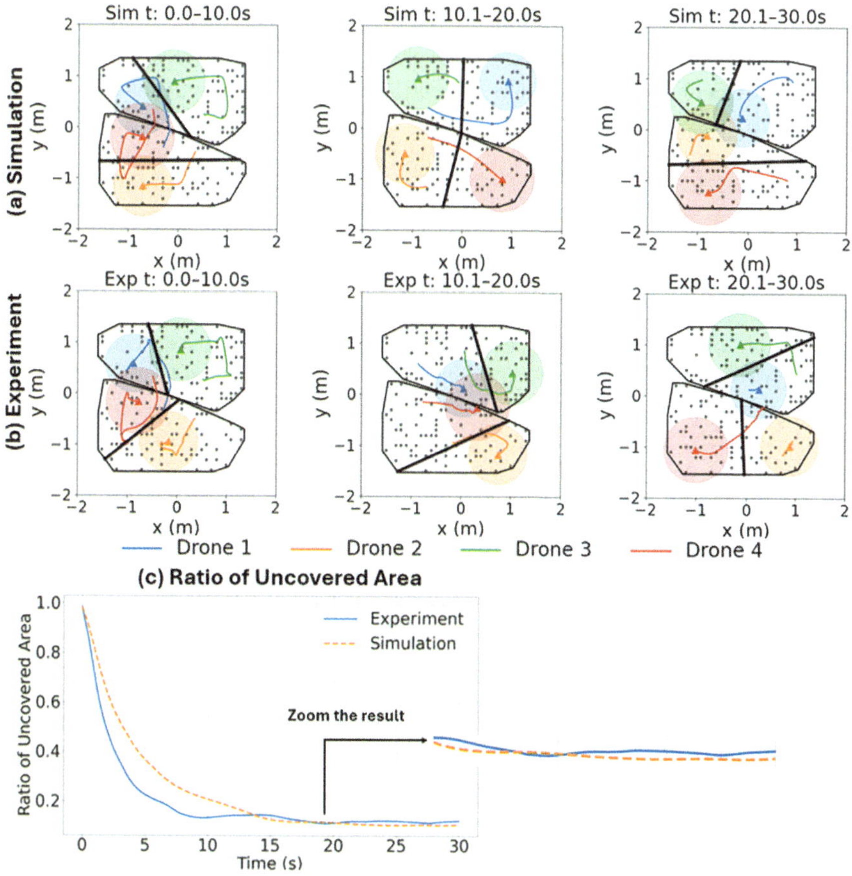

Fig. 5. Trajectories at interval t = 0–10 s, 10.1–20 s, and 20.1–30 s: **a** Simulation, **b** Experiment, and **c** Ratio of uncovered region versus time (0–30 s): experiment in blue, simulation in orange. Video of the experiment can be viewed at https://youtu.be/l0SbVZZOPSA

We categorize the coverage performance into two distinct phases: 0 – 15 s (phase 1) and 15 s – 30 s (phase 2). In Phase 1, both the simulation and experiment exhibit a decrease in the uncovered area ratio from approximately 0.9 to around 0.15. Interestingly, the experimental performance in Fig. 5 reaches the 0.15 level in 10 s, whereas the simulation requires approximately 15 s to achieve the same level. This discrepancy arises due to the difference in responses between drones in the simulation and the experiments. Specifically, the simulation assumes an idealized setting, where each velocity command translates instantly and precisely into drone movements without delay. In contrast, the real drones experience small latencies in wireless command transmission and control-loop processing, which momentarily interrupt drone velocity commands. During these brief communication delays, the physical drones continue moving forward slightly instead of stopping immediately. When drones receive the next velocity setpoint, they must quickly adjust their positions, resulting in more aggressive corrective movements. These sharper results in coverage performance in the experiment compared to the simulation, which was smoother due to delay-free trajectories.

In the experiment, we observe that the uncovered area ratio increases from 0.15 to approximately 0.17 between 11 and 15 s. Figure 6 highlights that the observed drone trajectories are associated with the performance drop. We assign a color to each drone based on its trail and sensing area shape: Drone 1 is blue, Drone 2 is green, Drone 3 is orange, and Drone 4 is red, in both simulation and experiment. During the 10–20 s interval, the experimental drones concentrate their search on the right side of the upper and lower cluster boundaries. In contrast, drones in the simulation spread evenly across both sides of each cluster, covering a greater number of target areas. Consequently, in the experiment, the left side of each cluster remained unmonitored, resulting in an increase in the uncovered area ratio to 0.17.

From 20 to 25 s and from 25.1 to 30 s, slight rebounds and oscillations appear in the experimental results. We observe that this pattern primarily occurs due to the minimal exploration activity of Drone 1 and Drone 2, whose trails in Fig. 5 are reduced compared to the interval between 10 and 20 s. By contrast, the simulation exhibits insignificant oscillations during these intervals because all four drones consistently continue to explore new targeted areas. The performance discrepancy between the simulation and experiment originates from how we implement the same single-integrator model in (15). In simulation, velocity commands u_i are executed at 50 Hz, as per the ideal setting, allowing the Voronoi-based coverage algorithm to run efficiently and result in smooth, coordinated drone motions. In contrast, the experiment is affected by real-world nonidealities, such as computational delay when updating the Voronoi coverage algorithm, communication delays when transmitting velocity commands over a radio link, and measurement delays when feeding the drone's motion updates from the Vicon motion-capture system. These factors may reduce the effective update rate below 50 Hz. Furthermore, although our high-level control assumes the Crazyflie behaves as a single integrator, the platform exhibits 6-DOF dynamics [15]. Consequently, in firmware, the same commands are realized through nested control loops, which introduce small delays and lags in the closed-loop response. Moreover, the unmodeled dynamics (rotor-wake interactions, aerodynamic drag) and actuator limitations (saturation, battery-voltage sag) further degrade the drones' motion, producing a non-smooth trajectory as each velocity command is applied with a small lag. Finally, the drones' mass and inertial limits how the real drones change the velocity and attitude, introducing an inertial lag that is absent in the single integrator dynamics used in simulation.

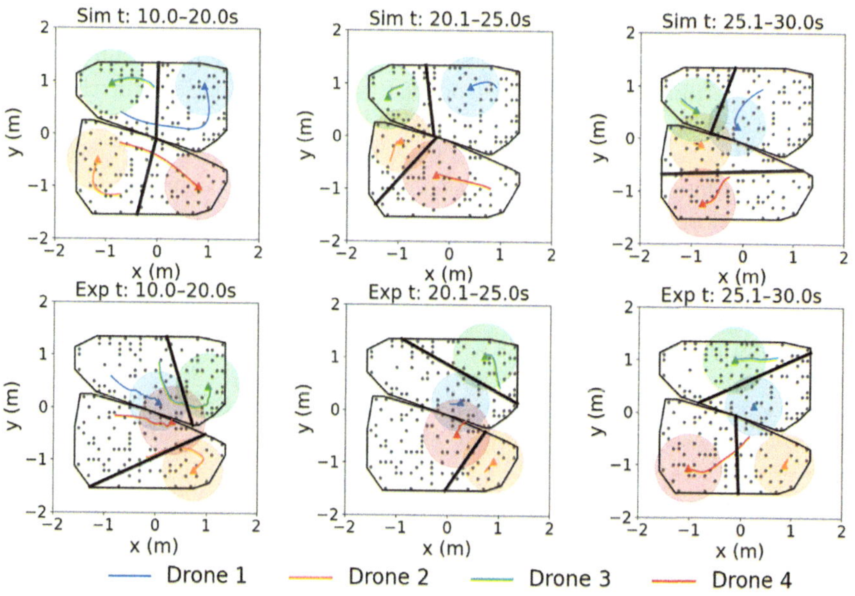

Fig. 6. This figure shows drones' movement within the time intervals, showing that this condition affects the coverage performance behavior

Other critical challenges to consider for implementation in a real application include the following. First, the TLCA assumes that each drone can determine its position in a global coordinate system, which can be achieved using technologies such as GNSS. Second, a ground station is required to track drone positions, update the density map, and send this information back to the drones. Lastly, drones need enough computational power and reliable communication systems with fast data transmission and sufficient bandwidth. One key challenge is that the control inputs for the persistent monitoring algorithm rely on calculations over Voronoi regions, approximated using discrete grids. This creates a trade-off between higher computational and communication requirements and the accuracy of the calculations.

6 Conclusion

In this study, we propose a Two-Layer Control Architecture (TLCA) for persistent monitoring with drone swarms, comprising task allocation, dynamic coverage, and safe monitoring. We formulate the task allocation as a multi-objective optimization that minimizes the uncovered area and unbalanced allocation in (3) and (4). We provide two equivalent solutions, a concave QP and an MILP, and evaluate their performance under varying β_f and β_g settings. Across Monte Carlo trials, we verified that $f(x)$ and $g(x)$ are in conflict in the sense of Definition 1, so no single objective dominates. Therefore, each objective yields its unique optimal solution. Computationally, the concave QP consistently outperformed the MILP in runtime across distributions and numbers of regions in Table 2, while obtaining the same optimal solutions. Building on these computational findings, we employ a concave QP in the experiment testbed with four Crazyflies to demonstrate the TLCA capability in persistent monitoring. Coupling Voronoi coverage control with our approach, we achieve reductions in the uncovered-area metric that closely follow simulation trends. During these experiments, we observed a discrepancy between simulation and experimental performance, mainly due to non-ideal experimental conditions compared to the idealized simulations. Despite this discrepancy, these findings show their effectiveness and practical applicability.

Acknowledgments. The authors wish to acknowledge the CSC—IT Center for Science, Finland, for providing computational resources, and the Tampere University Foundation for the travel grant.

References

1. Ferreira, J. F., Portugal, D., Andrada, M. E., Machado, P., Rocha, R. P., & Peixoto, P. (2023). Sensing and artificial perception for robots in precision forestry: a survey. *Robotics, 12*(5), 139.
2. Rabinovich, S., Curry, R. E., & Elkaim, G. H. (2018). Toward dynamic monitoring and suppressing uncertainty in wildfire by multiple unmanned air vehicle system. *Journal of Robotics, 2018*, 6892153, 12 pages.
3. Lappas, V., Shin,H. -S., Tsourdos, A., Lindgren, D., Bertrand, S., Marzat, J., Piet-Lahanier, H., Daramouskas, Y., & Kostopoulos, V. (2022). Autonomous unmanned heterogeneous vehicles for persistent monitoring. *Drones, 6*(94).
4. Alyassi, R., Khonji, M., Karapetyan, A., Chau, S. C. K., Elbassioni, K., & Tseng, C. M. (2022). Autonomous recharging and flight mission planning for battery-operated autonomous drones. *IEEE Transactions on Automation Science and Engineering, 20*(2), 1034–1046.
5. Ming, R., Jiang, R., Luo, H., Lai, T., Guo, E., & Zhou, Z. (2023). Comparative analysis of different uav swarm control methods on unmanned farms. *Agronomy, 13*(10), 2499.
6. Liu, Y., Liu, H., Tian, Y., & Sun, C. (2020). Reinforcement learning based two-level control framework of UAV swarm for cooperative persistent surveillance in an unknown urban area. *Aerospace Science and Technology, 98*, 105671.

7. Sargolzaei, A., Abbaspour, A., & Crane, C. D. (2020). Control of cooperative unmanned aerial vehicles: review of applications, challenges, and algorithms. In *Optimization, learning, and control for interdependent complex networks* (pp. 229–255).
8. Hübel, N., Hirche, S., Gusrialdi, A., Hatanaka, T., Fujita, M., & Sawodny, O. (2008). Coverage control with information decay in dynamic environments. *IFAC Proceedings Volumes, 41*(2), 4180–4185.
9. Gusrialdi, A., & Yu, C. (2014). Exploiting the use of information to improve coverage performance of robotic sensor networks. *IET Control Theory Application, 8*(13), 1270–1283.
10. Atman, M. W. S., Fikri, M. R., Priandana, K., & Gusrialdi, A. (2024). A Two-Layer Control Framework for Persistent Monitoring of a Large Area With a Robotic Sensor Network. *IEEE Access, 12*, 4153–4165.
11. Fikri, M. R., Atman, M. W. S., Nikulin, Y., & Gusrialdi, A. (2024). Efficient multi-robot task allocation with nonsmooth objective functions for persistent monitoring in large dispersed areas. In *2024 IEEE 20th International conference on automation science and engineering (CASE), Bari, Italy* (pp. 573–578).
12. Gusrialdi, A., Qu, Z., & Simaan, M. A. (2017). Distributed scheduling and cooperative control for charging of electric vehicles at highway service stations. *IEEE Transactions on Intelligent Transportation Systems, 18*(10), 2713–2727.
13. Preiss, J. A., Honig, W., Sukhatme, G. S., & Ayanian, N. (2017). Crazyswarm: a large nano-quadcopter swarm. In *2017 IEEE International conference on robotics and automation (ICRA), Singapore* (pp. 3299–3304).
14. Funada, R., Santos, M., Gencho, T., Yamauchi, J., Fujita, M., & Egerstedt, M. (2020). Visual coverage maintenance for quadcopters using nonsmooth barrier functions. In *2020 IEEE International conference on robotics and automation (ICRA), Paris, France* (pp. 3255–3261).
15. Bitcraze, A. B. (2024). Controllers in the Crazyflie. In Crazyflie-firmware documentation (Online). Available: https://www.bitcraze.io/documentation/repository/crazyflie-firmware/master/functional-areas/sensor-to-control/controllers/#cascaded-pid-controller. Accessed: 08 August 2025.

BLE for the Cooperative Awareness of Miniature UAVs: An Experimental Analysis of the Different Advertising Modes

Sandaruwan Jayaweera(✉), Matti Hämäläinen, and Konstantin Mikhaylov

Centre for Wireless Communications, University of Oulu, Oulu, Finland
{sandaruwan.jayaweera,matti.hamalainen,
konstantin.mikhaylov}@oulu.fi

Abstract. Vehicle-to-Vehicle (V2V) Cooperative Awareness (CA) is an Intelligent Transport Systems (ITSs) technology standardized by the European Telecommunications Standards Institute (ETSI). It is designed to ensure future transportation systems are safe, reliable, and efficient. Miniature Unmanned Aerial Vehicles (UAVs) have been increasingly utilized by various applications due to their expanding functionality. Consequently, a UAV-to-UAV CA solution is anticipated to become necessary in the near future. However, the implementation of a UAV-CA using Vehicle-to-Everything (V2X) communication technologies such as IEEE802.11p, IEEE802.11bd, and C-V2X is impossible on miniature UAVs due to overly high power consumption, the size and weight of commercial On-Board-Units (OBUs). This paper advocates using Bluetooth Low Energy (BLE) as a carrier for Cooperative Awareness Messages (CAM), benefiting from the low cost and low power consumption characteristics of this technology. Specifically, the paper reports the proof-of-concept implementations of CAM tailor-made for UAVs on top of BLE4 and BLE5 stacks and demonstrates the range and power consumption performance obtained using analytical and experimental approaches. The results indicate that BLE is able to provide multiple low-power solutions with adequate range capabilities while meeting traffic requirements of CAM.

Keywords: Vehicle-to-vehicle · Vehicle-to-everything · Unmanned aerial vehicles · Intelligent transportation systems · Bluetooth low energy

1 Introduction

The need for Cooperative Awareness (CA) for Unmanned Aerial Vehicles (UAVs) has become evident as their deployment in various applications continues to expand. CA, being a cornerstone technology for Intelligent Transport Systems (ITSs), enables real-time information sharing among system participants to

M. Laakkonen (ed.), *DroneAI for Resilient Nature*,
https://doi.org/10.1007/978-3-032-19008-6_6

enhance situational awareness and ensure operational safety [1]. Without access to CA, UAVs within an ITS face limitations in avoiding collisions, maintaining formation in swarm operations, and achieving effective coordination with other aerial or ground entities [2]. The increasing density of UAV operations in shared airspace has further underscored the importance of equipping them with CA capabilities to reduce the risk of accidents and improve overall efficiency. Through CA, UAVs can be provided with timely and accurate data on their surroundings, enabling safer navigation, better decision-making, and the potential to contribute to cooperative missions with other ITS components.

Furthermore, the advantages of CA enabled by intercommunication between UAVs and ground vehicles are significant. Enhanced situational awareness is provided to all participants, allowing for improved safety through collision avoidance and coordinated navigation. Real-time information exchange facilitates traffic flow optimization and supports critical applications, such as emergency response and road hazard detection. Additionally, the integration of UAVs with ground vehicles ensures efficient utilization of shared air and ground spaces, enabling synergistic operations and reducing delays in transportation systems. This interconnected framework contributes to the overall efficiency and safety of ITS, as well as humans.

CA service utilizes wireless communication to exchange information between road users and roadside infrastructure, thereby enhancing traffic efficiency and road safety. To enable these capabilities, the Cooperative Awareness Message (CAM) protocol specification has been defined by the European Telecommunications Standards Institute (ETSI) [3]. With the increasing adoption of UAVs, particularly miniature UAVs, valued for their affordability and ability to access areas unreachable by ground vehicles, the integration of UAVs into ITS is recognized as essential. However, support for aerial traversal is not provided by the existing CAM specification. To overcome this limitation, a modified UAV-CAM specification has been introduced to extend CAM functionality to UAV operations by Jayaweera et al. in [4]. This UAV-CAM is designed by exploiting the flexibilities of the ETSI CAM specification, which allows extending the protocols to support future ITS applications [3]. Therefore, a CAM parser can be easily modified to parse UAV-CAM. This interoperability is crucial for ensuring seamless integration of UAV-CAM within existing ITS frameworks, enhancing communication efficiency. A possible scenario where UAV-CAM coexists with conventional CAM in an urban ITS is depicted in Fig. 1. In Fig. 1, RSU stands for a roadside unit.

The communication technology suitable for UAV-CAM is determined by the data payload size it supports, operational requirements, and power source capacity. Additionally, a balance must be achieved between the robustness and reliability of UAV-CAM delivery and resource efficiency. In this work, the feasibility of UAV-CAM for real-life UAVs using Bluetooth Low Energy (BLE) communication technology is explored. BLE is selected due to its widespread availability, low power consumption, and compact device size. Furthermore, BLE technology is already supported by smartphones and indoor localization systems, providing

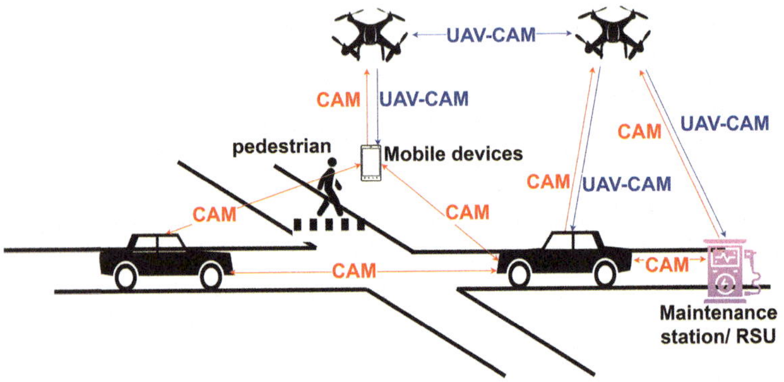

Fig. 1. Conceptualized UAV-CAM coexistence alongside conventional CAM

further visibility and additional advantages for UAVs, respectively. The advertising feature of BLE is utilized to accommodate the broadcasting requirements of UAV-CAM. Additionally, flexibility in creating various UAV-CAM solutions for miniature UAVs, depending on specific scenarios, is offered by the different physical layers and advertising modes of BLE.

In the following sections, the paper starts by providing a brief review of related work on UAV CA and BLE advertising. Then, the technical details of BLE advertising are presented. Next, the test setup and the experimental procedure are explained. The experimental results are presented alongside the analytical results in the next section. Finally, the paper concludes by highlighting potential avenues for future research.

2 Related Work

Two main categories of prior research are considered in this work. First, the cooperation and CA techniques within UAV operations are examined. The improvement of control, safety, and maneuverability in multi-UAV networks through CA has been demonstrated in several studies. For instance, the cooperative sharing of sensor information among multiple UAVs to enhance Global Positioning System (GPS)-based localization is suggested by Lee in [5]. A Situational Awareness (SA) scheme leveraging multiple sensors within a cooperative multi-UAV network is proposed by Liao et al. in [6]. The complexity introduced by cooperation and collaboration in multi-UAV networks has necessitated optimization in control and management [7]. Consequently, some studies have focused on optimizing resource allocation and task distribution while maintaining cooperation within UAV networks. Under this scope, Wang et al. in [8] propose an efficient gateway selection mechanism for cooperative miniature UAVs, and Gu et al. in [9] propose a cooperative task assignment scheme based on swarm intelligence for multi-UAV surveillance systems. However, most of these studies focus solely on improving specific aspects of multi-UAV operations through cooperation or collaboration, sharing only a limited and specific set of information. To achieve proper CA, a more extensive set of information must be shared among UAVs. UAV-CAM has been identified as a suitable solution, as it specifies a comprehensive set of data fields to be shared among UAVs, which can support applications such as localization, collision avoidance, traffic management, path planning, and SA [4].

Second, the applications and performance of BLE advertising are analyzed. BLE advertising, often referred to as BLE beacons, has typically been applied for navigation or localization purposes. For instance, a system utilizing BLE beacons to locate lost children and achieve a 60% success rate is proposed by Ito et al. in [10]. The multiple advertising modes of BLE, each with distinct capabilities, are analyzed in an indoor environment by Badihi et al. in [11]. In UAV-specific scenarios, BLE beacons have been applied for navigation and positioning [12, 13]. These studies have demonstrated promising outcomes for BLE advertising in localization and navigation tasks. Building upon this foundation, this study advocates the use of BLE to enable CA for miniature UAVs and explores the practical performance of exemplary implementation concerning communication range and energy consumption.

3 Methodology

In this work, a simple and cost-effective solution is proposed and investigated, leveraging BLE, which is among the most widely adopted IoT connectivity technologies in terms of supported connections. Additionally, to ensure efficient CA and allow sufficient reaction time for UAVs, the UAV-CAM messages must be delivered reliably and over extended distances. Accordingly, this section establishes the foundation of the analysis by evaluating the probability of UAV-CAM packets being successfully delivered or lost as a function of distance.

3.1 Cooperative Awareness for Unmanned Aerial Vehicles

The basic concept of the UAV-CAM was previously presented in [4], and the general structure and size information are summarized to ensure clarity and completeness in this paper. Therefore, a high-level structure of the main containers and headers in UAV-CAM is depicted in Fig. 2. It should be noted that location information is included within the first three containers (28 bytes). Therefore, 28 bytes of UAV-CAM are considered sufficient for most basic miniature UAV applications. However, when all five containers are utilized, the size can range from 79 to 948 bytes. For instance, when the first three containers (28 bytes), the HF container (36 bytes), the UAV container's safety area radius (2 bytes), the UAV container length headers (2 bytes), and 16 pathpoints (each 11 bytes) are included, the total packet size increases to 244 bytes after applying byte stuffing [4].

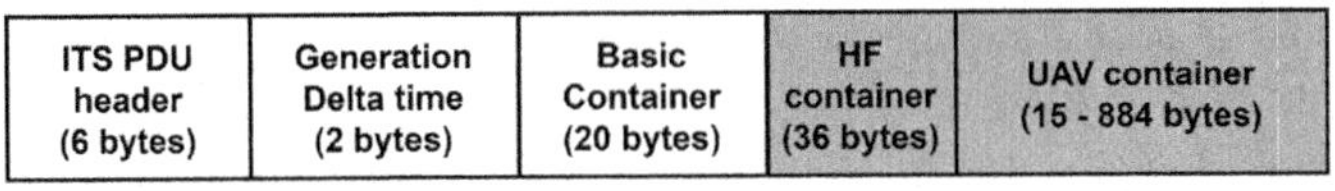

Fig. 2. High-level structure of the UAV CAM

3.2 Bluetooth Low Energy

BLE operates within the unlicensed 2.4 GHz Industrial, Scientific, and Medical (ISM) band, utilizing binary frequency modulation to reduce transceiver complexity. BLE operates over 40 frequency channels in the 2.4–2.4835 GHz range, with each channel occupying 1 MHz bandwidth and spaced 2 MHz apart. Among these, three channels (channels 37, 38, and 39) are designated as primary advertising channels, while the remaining 37 channels are general-purpose channels traditionally allocated for data transfer after a connection is established, as defined by the Bluetooth 4.2 specification [14]. These data channels can also be utilized as secondary advertising channels for extended advertising, as specified in Bluetooth 5.0 [15] and later versions.

Legacy Advertising Advertising modes prior to Bluetooth 5 are referred to as legacy advertising in this work. Within legacy advertising, the Protocol Data Unit (PDU) permits a maximum Advertising Data (AD) size of 31 bytes. Although multiple AD types are supported, only a single AD type (i.e., the Manufacturer-specific data) can encapsulate a reduced version of the 28-byte UAV-CAM within the AD data field.

Extended Advertising Although Bluetooth 4.2 increased the maximum PDU size for connected-mode BLE packets to 259 bytes, the limitation of a 39-byte PDU size for advertisements remained unchanged. However, with the introduction of Bluetooth 5, advertisements utilizing secondary advertising channels for larger payloads through extended advertising were enabled. Specifically, Bluetooth 5 allows the use of a primary advertisement to transmit a pointer to a secondary advertisement, which can accommodate up to a 257-byte PDU. With a packet structure similar to legacy, extended advertising is able to host 244 byte UAV-CAM within AD data, considering data overheads.

Bluetooth 5 uses 3 physical layers (PHYs) for BLE extended advertising [15]. They are:

- 1M PHY: This physical layer, originally defined in Bluetooth versions preceding Bluetooth 5, remains mandatory for all BLE devices. Operating with a symbol rate of 1 Msym/s, it supports a data rate of 1 Mbps using Gaussian Frequency Shift Keying (GFSK) modulation. This PHY is utilized for both primary and secondary advertising.
- 2M PHY: This PHY is an optional configuration designed to support high-throughput applications and is exclusively utilized in secondary advertising. Operating at a symbol rate of 2 Msym/s, it achieves a data rate of 2 Mbps using GFSK modulation. However, the range provided by the 2M PHY is reduced compared to the 1M PHY, and the higher data rate increases its susceptibility to inter-symbol interference (ISI) [16]. This heightened ISI arises due to the shorter symbol duration, making the 2M PHY more prone to multipath delays that result in overlapping symbols.
- Coded PHY: This optional PHY is designed to enhance the range and accuracy of BLE transmission while reducing the data rate. The two available Coded PHYs support data rates of 500 kbps and 125 kbps, utilizing Forward Error Correction (FEC) schemes, referred to as $S = 2$ and $S = 8$, respectively. In the $S = 2$ Coded PHY, 1 bit is represented by 2 symbols, with a symbol rate of 1 Msym/s, while in the $S = 8$ Coded PHY, 1 bit is represented by 8 symbols, resulting in the effective data rates mentioned above. The FEC used in BLE Coded PHY is based on a convolutional coding scheme, which introduces redundancy by encoding each bit into multiple output symbols to improve robustness against errors. The Coded PHY uses GFSK modulation, the same as other BLE PHYs. The Coded PHY can be employed for both primary and secondary advertising. In this paper, the Coded PHY with the $S = 8$ FEC scheme has been analyzed.

It should be noted that in what follows, unless otherwise specified, the use of the 1M, 2M, and Coded refers to 1M PHY extended advertising, 2M PHY extended advertising, and Coded PHY extended advertising, respectively.

3.3 Analytical Models

To determine the variation of Packet Error Rate (PER) with communication distance, pathloss must be obtained. A Two Ray (TR) ground reflection pathloss model has been derived in [17] as,

$$\mathrm{PL[dB]} = 40\log_{10}(d) - 10\log_{10}(h_r^2 h_t^2), \tag{1}$$

where PL[dB] represents the path loss in dB, and d denotes the link distance. Considering that both the transmitter and receiver are positioned 1 m above the ground, the transmitter and receiver heights are set as $h_t = h_r = 1$ m. Therefore, PL[dB] is used to calculate the Received Signal Strength (RSS) at the receiver, as given by,

$$\text{RSS}_{rx} = P_{tx} - \text{PL}[\text{d}B] + G_{tx} + G_{rx} - P_{\text{sen},rx}, \tag{2}$$

where the transmit power (P_{tx}) is set to 0 dBm, the transmitter antenna gain (G_{tx}) is 0 dBi, and the receiver antenna gain (G_{rx}) is 0 dBi. The analysis is conducted for all advertising modes. The experiment was carried out using Nordic Semiconductor nRF52840 dongles. Therefore, corresponding receiver sensitivity values from the nRF52840 product specification are used in the analysis to enable a more accurate comparison with the experimental results. Accordingly, based on the radio specifications from Nordic Semiconductor, the receiver sensitivity ($P_{\text{sen},rx}$) is set to −93 dBm for 1 M and legacy, −89 dBm for 2 M, and −103 dBm for Coded [18]. It should be noted that practical receiver sensitivities have been found to be significantly better than the sensitivities listed in the Bluetooth specification [19].

The symbol error rate (SER) can be calculated as specified in [20] by,

$$\text{SER} = \frac{1}{2}\exp(-E_s/2N_0), \tag{3}$$

where N_0 is the noise power spectral density and E_s is the energy per symbol. (3) can be further simplified by, $E_s/N_0 = E_b/N_0 = B_{\text{DS}}/R_b \times \text{SNR}_{rx}$ where E_b is the energy per bit and SNR_{rx} is the Signal-to-Noise ratio at the receiver [21]. Note that redundancy bits of Coded PHY are considered as information bits in this analysis. Therefore, according to [21], this results in (3) being reduced to,

$$\text{SER} = \frac{1}{2}\exp(-B_{\text{DS}}/R_b \times \text{SNR}_{rx}/2), \tag{4}$$

where B_{DS} is the bandwidth of the receiver filter and R_b is the data rate. Furthermore, with a zero-interference scenario assumed, $\text{SNR}_{rx} = \text{RSS}_{rx}$. This transforms (4) into,

$$\text{SER} = \frac{1}{2}\exp(-B_{\text{DS}}/R_b \times \text{RSS}_{rx}/2). \tag{5}$$

The RSS value in dB obtained in (2) should be converted to the linear ratio (non-logarithmic) before being used in (5). In this research, BLE uses GFSK modulation with $B_{\text{DS}} = 1\,\text{MHz}$. The data rate, $R_b = 1$ Mb/s for legacy, 1M, and Coded while $R_b = 2$ Mb/s for 2M.

The AD data size of 28 bytes results in packet lengths $k = 46 \times 8$ for legacy advertisements. Similarly, an AD data size of 244 bytes results in $k = 260 \times 8$ for secondary advertisements in 1 and 2M. The packet length $k = 23 \times 8$ for primary advertisements in 1 and 2M. Due to $S = 8$ FEC scheme, the packet sizes will be approximately 8 times larger for Coded advertisements. Further details on BLE packet structures can be found in the Bluetooth Specification, available only from Bluetooth 5.0 onwards [15]. With the assumption of SER being the same for every symbol, the PER calculation by [22] can be reduced to,

$$\text{PER} = 1 - (1 - \text{SER})^k. \tag{6}$$

The secondary advertisement is preceded by three primary advertisements transmitted over separate advertising channels. With the PER of a single primary advertisement denoted as PER_p and PER of secondary advertising denoted as PER_s the PER of extended advertising (PER_{ext}) can be calculated as,

$$\text{PER}_{\text{ext}} = 1 - (1 - \text{PER}_p^3)(1 - \text{PER}_s). \tag{7}$$

4 Experimental Setup

To obtain PER of the proof-of-concept implementation, an outdoor experiment has been carried out on the road depicted in Fig. 3a [23]. This experimental setup will be explained in this section.

In this experiment, two nRF52840 dongles, one Nordic power profiler kit II (PPK II), and an nRF52840 development kit (DK) were used, as depicted in Fig. 3b. Two tripods were used to position two dongles 1 m above the ground. Transmitter (Tx) dongle was configured to transmit at 0 dBm power, while the other functioned as the Receiver (Rx).

The components used in this experiment were selected due to their cost, commercial availability, reconfigurability, size, and performance. The implementation was conducted entirely using nRF52840 dongles, which are suitable for mounting on small UAVs. The nRF52840 system on chip (SoC) on the dongle was programmed utilizing the nRF Programmer v3.0.7. The program was developed and implemented using the nRF Connect SDK (NCS) v2.3.0 within a Visual Studio (VS) Code environment. On Tx, the advertising interval was configured to 95 ms and 995 ms to achieve UAV-CAM frequencies of 10 Hz and 1 Hz, respectively. Furthermore, the Tx was configured to enter idle mode between advertising events. On the scanner side, both the scan window and the scan interval were configured to 30 ms.

The standard nRF52840 dongles are configured by default to be powered via the USB interface. For our setup, they were reconfigured to draw power through the connector pins in order to interface with the PPK II [24,25]. The reconfigured transmitting dongle was powered through the PPK II during UAV-CAM transmission, as depicted in Fig. 3b.

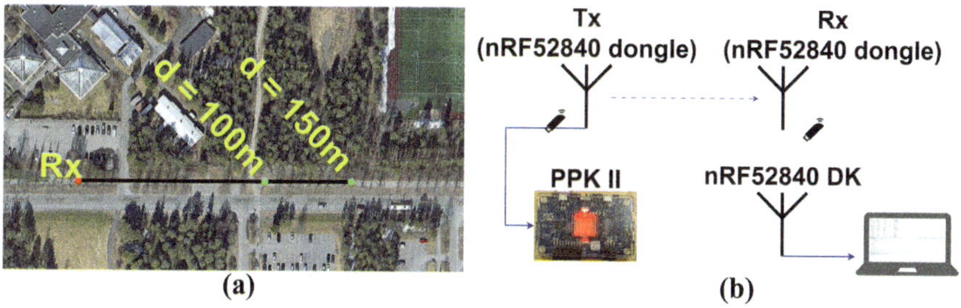

Fig. 3. **a** The road used for the experiment. **b** The experimental test setup

5 Evaluation Results and Discussion

This section presents our results for range and power consumption to support our claims. The range of BLE-based UAV-CAM is analyzed by plotting the analytical models for PER from Sect. 3.2 along with the experimental results from Sect. 4. The power consumption of the UAV-CAM transmission was measured using the PPK II.

5.1 Packet Error Rate Analysis

The variation in PER with communication distance is depicted in Fig. 4 for both the experiment (Exp.) and the four analytical models described in Sect. 3.2. Communication ranges of approximately 100 m were achieved by both legacy and Coded advertising. Additionally, communication ranges of around 75 m and 70 m were achieved by 1M and 2M advertising, respectively.

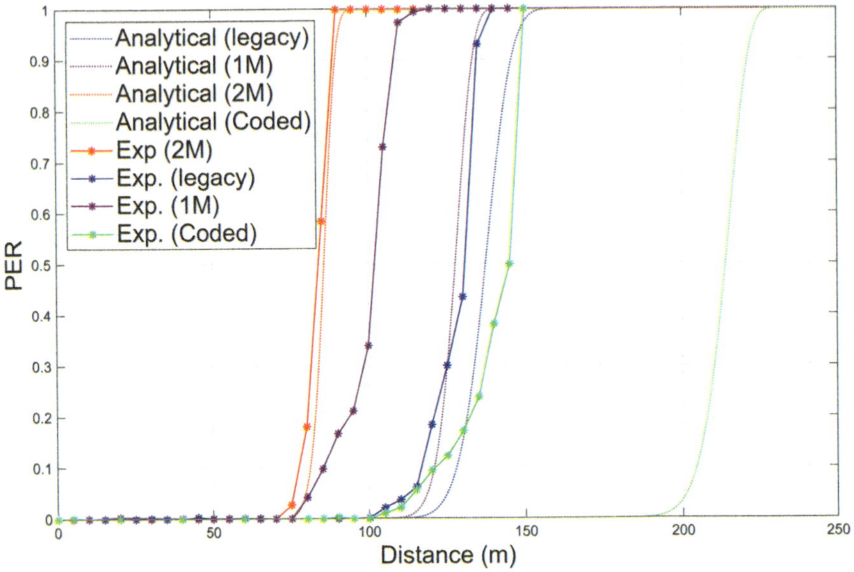

Fig. 4. PER against the distance of different BLE PHYs with UAV-CAM

A significant disparity was noted between the experimental and analytical results for Coded advertising. A communication range of nearly 200 m was predicted by the analytical models, and this exceeds twice the range observed experimentally. For 1M advertising, an analytical range of approximately 110 m was calculated, which was notably higher than the 75 m observed experimentally. Better match was observed between analytical and experimental results for legacy advertising, with an analytical range of approximately 120 m. The minimal disparity was noted for 2M advertising, with the analytical range being 75 m and the experimental range being 70 m.

Overall, the experimental ranges of UAV-CAM for both legacy and extended advertising were found to be lower than those obtained through analytical methods. This difference was partially attributed to interference, which was not controlled during the experiment. However, error-free communication was consistently observed within a 70 m range under practical conditions.

This conclusion is particularly significant, as it indicates that two drones traveling towards each other at the maximum speed allowed by EU regulations (19 m/s) [26] would have approximately 2 s to exchange 20 packets, enabling mutual detection and collision avoidance.

5.2 Power Consumption Analysis

The power consumption of different BLE advertising modes for 1–10 Hz UAV-CAM report frequencies is illustrated in Fig. 5. A linear relationship can be observed between the power consumption and the report frequency. Due to the increased effect of the random delay in BLE advertising as the advertising interval decreases, the highest UAV-CAM frequencies have a slight deviation from intended integer values. Throughout the experiment, the power consumption of Coded was significantly higher than other advertising modes. This variation in power consumption across PHYs is primarily attributed to the radio-on time required to transmit a UAV-CAM packet. Primary advertising incurs the lowest power usage, as it uses the 1 Mbps data rate with a relatively short packet size. The 2M PHY follows, benefiting from its 2 Mbps data rate, which reduces transmission time. The 1M PHY, although using the same packet size as 2M PHY, consumes more power due to its slower 1 Mbps data rate. Coded PHY results in the highest power consumption because it combines a 1 Mbps data rate with a significantly larger packet size, leading to extended transmission durations.

For the lowest possible UAV-CAM rate of 1 Hz, the average power consumption of legacy advertising has been around 0.05 mW, while a power consumption of around 0.22 mW can be observed at a frequency of around 10 Hz. Similarly, a power consumption of 0.13 mW, 0.15 mW, 0.45 mW, for 1 Hz, and a power consumption of around 0.45 mW, 0.66 mW, 3.9 mW for 10 Hz can be observed for 2M, 1M and Coded respectively. When compared to the power consumption of a miniature UAV, which ranges from approximately 8 to 300 W, these power consumptions are significantly (over three orders of magnitude) lower [27–29]. Furthermore, considering that a miniature lithium-ion battery with a capacity of 400 mAh is used for advertisement, it enables [30]

- around 300 h for the most power hungry 10 Hz Coded advertising;
- around 24000 h for the least power hungry 1 Hz legacy advertising.

6 Conclusion

Miniature UAVs are starting to play a more important role these days. This paper has delved into an extension of the existing CAM standard tailor-made

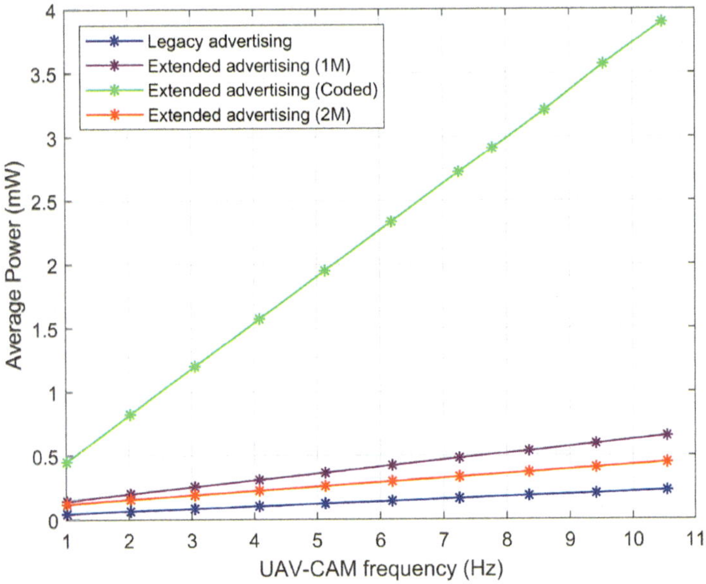

Fig. 5. Power consumption of nRF52840 dongle implementation of UAV-CAM

to support UAV CA. A UAV-CAM implementation has been reported, and its performance has been analyzed using analytical and experimental methods. The Coded and legacy advertising-based UAV-CAM has been able to achieve a range of 110 m, while 75 m and 70 m ranges were achieved by 1M and 2M. Furthermore, even though legacy advertising provides a limited version of CA, due to its significantly lower energy consumption, it can become a viable CA solution in power-saving scenarios.

To improve the accuracy and relevance of the analytical models, future work will incorporate more realistic channel conditions, including multipath fading, noise, interference, and frequency-selective effects, as well as explore the use of Channel State Information (CSI) for enhanced channel characterization. Furthermore, modifications to the Decentralized Environmental Notification Message (DENM) structure will be introduced to facilitate event-triggered UAV applications. Since DENM are event-triggered, they can be used for such use cases as emergency docking, refueling, landing, and collaborative operations. The range can be further improved by increasing the transmission power. To address the limitations observed in BLE range and reliability, especially for fast-moving UAVs, future work will also explore alternative communication technologies such as IEEE802.11p/bd, LTE-V2X, and NR-V2X. Additionally, hybrid communication architectures combining multiple BLE PHYs or technologies will be investigated to optimize the trade-offs between power efficiency, scalability, and real-time performance in diverse UAV operational scenarios. These aspects will be further analyzed in future publications.

Acknowledgments. This work is partially funded by the Research Council of Finland via projects RoboMesh (grant 336060), FIREMAN (2430407411), MRATSafeDrone (grant 341111), 6G Flagship (grant 369116), and University of Oulu Graduate School.

References

1. Wang, Y., Fan, Y., Wang, G., Qiao, S., & Feng, R. (2022). Research on cooperative situational awareness of unmanned surface vehicle swarm based on multi-radar joint detection. In *7th International conference on control, automation, robotics and vision engineering (CACRE)* (pp. 51–57). https://doi.org/10.1109/CACRE54574.2022.9834192.
2. Huang, S., Zhang, H., & Huang, Z. (2024). E2CoPre: Energy efficient and cooperative collision avoidance for UAV swarms with trajectory prediction. *The IEEE Transactions on Intelligent Transportation Systems, 25*(7), 6951–6963. https://doi.org/10.1109/TITS.2023.3342161
3. European Telecommunications Standards Institute (2019). Intelligent transport systems (ITS); Vehicular communications; Basic set of applications; Part 2: Specification of cooperative awareness basic service. ETSI EN 302 637-2 V1.4.1.
4. Jayaweera, S., Mikhaylov, K., & Hämäläinen, M. (2022). Enabling cooperative awareness for UAVs: ETSI CAM protocol extension. In *Proceedings of joint European conference on networks and communications & 6G Summit (EuCNC/6G Summit)* (pp. 339–344). https://doi.org/10.1109/EuCNC/6GSummit54941.2022.9815616
5. Lee, C. Y. (2020). Cooperative drone positioning measuring in internet-of-drones. In *IEEE 17th Annual consumer communications & networking conference (CCNC)* (pp. 1–3). https://doi.org/10.1109/CCNC46108.2020.9045111
6. Liao, Z., Wang, S., Shi, J., Sun, Z., Zhang, Y., & Sial, M. B. (2023). Cooperative situational awareness of multi-UAV system based on improved D-S evidence theory. *Aerospace Science and Technology, 142*, 108605. ISSN 1270-9638. https://doi.org/10.1016/j.ast.2023.108605
7. Saffre, F., Karvonen, H., Hildmann, H. (2025, January). The why and how of polymorphic artificial autonomous swarms. *Drones, 9*(1). ISSN 2504-446X. https://doi.org/10.3390/drones9010053
8. Wang, J., Jiang, C., Han, Z., Ren, Y., Maunder, R. G., & Hanzo, L. (2017). Taking drones to the next level: cooperative distributed unmanned-aerial-vehicular networks for small and mini drones. *The IEEE Vehicular Technology Magazine, 12*(3), 73–82. https://doi.org/10.1109/MVT.2016.2645481
9. Gu, J., Su, T., Wang, Q., Du, X., & Guizani, M. (2018). Multiple moving targets surveillance based on a cooperative network for multi-UAV. *IEEE Communications Magazine, 56*(4), 82–89. https://doi.org/10.1109/MCOM.2018.1700422
10. Ito, A., Xiang, F., Hatano, H., & Uchiyama, H. (2018). A study on observing a lost child using BLE beacons. In *Sixth international symposium on computer networks workshop (CANDARW)* (pp. 67–70). https://doi.org/10.1109/CANDARW.2018.00021
11. Badihi, B., Sheikh, M.U., Ruttik, K., & Jäntti, R. (2020). On performance evaluation of BLE 5 in indoor environment: an experimental study. In *The IEEE 31st annual international symposium on personal, indoor and mobile radio communications* (pp. 1–5). https://doi.org/10.1109/PIMRC48278.2020.9217132

12. Yukhimets, D. A., Naidanova, I. E., & Andrukhovsky, N. V. (2024). Local navigation system for a robot based on a network of BLE beacons. In *The international Russian automation conference (RusAutoCon)* (pp. 611–616). https://doi.org/10.1109/RusAutoCon61949.2024.10694049
13. Ariante, G., Ponte, S., & Del Core, G. (2022). Bluetooth low energy based technology for small UAS indoor positioning. In *IEEE 9th International Workshop on Metrology for AeroSpace (MetroAeroSpace)* (pp. 113–118). https://doi.org/10.1109/MetroAeroSpace54187.2022.9856321
14. Bluetooth Special Interest Group. Bluetooth Core Specification v4.2 (2014).
15. Bluetooth Special Interest Group. Bluetooth Core Specification v5.0 (2016).
16. Muhammad, S., Kalaa, M. O. A., & Refai, H. H. (2021). Wireless coexistence of cellular LBT systems and BLE 5. *IEEE Access, 9*, 24604–24615. https://doi.org/10.1109/ACCESS.2021.3056909
17. Rappaport, T. S. (1996). *Wireless communications: principles and practice.* IEEE Press. ISBN 0780311671.
18. Nordic semiconductor, N. (2024). Receiver sensitivity of nRF52840 radio. https://docs.nordicsemi.com/bundle/ps_nrf52840/page/radio.html. Accessed: 14 October 2024.
19. Böcker, S., Arendt, C., & Wietfeld, C. (2017). On the suitability of Bluetooth 5 for the internet of things: performance and scalability analysis. In *The 28th annual IEEE international symposium on personal, indoor and mobile radio communications* (pp. 1–7). https://doi.org/10.1109/PIMRC.2017.8292720
20. IEEE P802.15. Coexistence analysis of IEEE Std 802.15.4 with other IEEE standards and proposed standards (2010, September).
21. Natarajan, R., Zand, P., & Nabi, M. (2016). Analysis of coexistence between IEEE 802.15.4, BLE and IEEE 802.11 in the 2.4 GHz ISM band. In *Proceedings of 42nd Annual Conference of the IEEE Industrial Electronics Society (IECON)* (pp. 6025–6032). https://doi.org/10.1109/IECON.2016.7793984
22. Karvonen, H., Mikhaylov, K., Acharya, D., & Rahman, M. M. (2020). Performance evaluation of Bluetooth low energy technology under interference. In C. Sugimoto, H. Farhadi, M. Hämäläinen (Eds.), *13th EAI international conference on body area networks* (pp. 147–156). Springer.
23. Google (2024). Google Maps location of the road park used for the experiment. https://maps.app.goo.gl/RfD8bhPgo6ZBdnDdA. Accessed: 22 August 2024.
24. Nordic semiconductor. External regulated source for nRF52840 dongle (2025). https://docs.nordicsemi.com/bundle/ug_nrf52840_dongle/page/UG/nrf52840_Dongle/hw_power_ext_reg_source.html. Accessed: 13 Febuary 2025.
25. Nordic semiconductor. Power profiler Kit II (2025). https://docs.nordicsemi.com/bundle/ug_ppk2/page/UG/ppk/PPK_user_guide_Intro.html. Accessed: 27 May 2025.
26. European Union. Commission implementing regulation (EU) 2020/639 of 12 May 2020 amending implementing regulation (EU) 2019/947 as regards standard scenarios for operations executed in or beyond the visual line of sight. *The Official Journal of the European Union, 63*, 1–31. ISSN 1977-0677. http://data.europa.eu/eli/reg_impl/2020/639/oj
27. Olejnik, D.A., Wang, S., Dupeyroux, J., Stroobants, S., Karasek, M., De Wagter, C., & de Croon, G. (2022). An experimental study of wind resistance and power consumption in MAVs with a low-speed multi-fan wind system. In *International conference on robotics and automation (ICRA)* (pp. 2989–2994). https://doi.org/10.1109/ICRA46639.2022.9811834

28. Ganchaudhuri, P., Bhawal, C. (2023). Power consumption of a quadrotor based on maneuvers. In *IEEE Guwahati subsection conference (GCON)* (pp. 01–06). https://doi.org/10.1109/GCON58516.2023.10183455
29. Cabuk, U. C., Tosun, M., Dagdeviren, O., & Ozturk, Y. (2024). Modeling energy consumption of small drones for swarm missions. *The IEEE Transactions on Intelligent Transportation Systems, 25*(8), 10176–10189. https://doi.org/10.1109/TITS.2024.3350042
30. Data Power Technology Limited (2015). Product specifications—polymer Li-ion rechargeable battery. https://cdn.sparkfun.com/datasheets/Prototyping/spe-00-502535-400mah-en-1.0ver.pdf. Accessed: 14 October 2024.

Self-Organising Drone Swarms for Detection and Monitoring of Wildfires

Sebastian Coffeng[1], Hanno Hildmann[2,3], Hannu Karvonen[1(✉)], and Fabrice Saffre[1]

[1] VTT, Espoo, Finland
{sebastian.coffeng,hannu.karvonen,fabrice.saffre}@vtt.fi
[2] TNO, Den Haag, AK, The Netherlands
hanno.hildmann@tno.nl
[3] HHS, Smart Sensor Systems Lab, Den Haag, EP, The Netherlands
https://www.vttresearch.com/

Abstract. This chapter investigates the use of multiple unmanned platforms (a swarm) for environmental sensing. The use-case is that of wildfire detection and monitoring. The contribution of this chapter is theoretical in nature, the results are collected from simulations and numerical experiments. Specifically, the impact of varying communication ranges is explored with regard to the individual performances for the search- and the monitoring-task as well as the use of communication range as a control parameter for the trade-off between exploration and exploitation.

Keywords: Autonomous drone swarms · Self-organization · Detection · Environmental monitoring · Wildfire · Swarm intelligence · Decentralization · Search algorithms · Local information sharing

1 Introduction

There are currently no human operated systems that facilitate a continuous and pervasive monitoring of the vast Finnish forests. Autonomous drone swarms are increasingly considered a cost-effective alternative to manual human work to perform a variety of environmental monitoring tasks such as this. These types of tasks often involve maintaining large and remote wilderness areas, many of which are not easily accessible due to a lack of infrastructure, under constant observation. The (early) detection, monitoring, and possible containment of wildfires is one such example from a wide range of civilian and military applications.

To operate efficiently without human guidance for extended periods and to respond adequately to unpredictable events within their assigned territory, drone swarms must be able to leverage collective intelligence principles and maximize cooperative effects. Self-organization is known to play a key role in the emergence of such adaptive responses in nature. The underlying decision-making functions that have evolved to improve the fitness of (particularly) social insects can be modified or augmented to meet artificial mission requirements.

M. Laakkonen (ed.), *DroneAI for Resilient Nature*,
https://doi.org/10.1007/978-3-032-19008-6_7

1.1 What This Chapter Is About

In this chapter, we describe such a set of simple navigation decision rules that, when followed concurrently by all members of the swarm, can prevent or at least reduce duplication of effort in the surveillance phase of a wildfire detection and monitoring mission. We present the results of a systematic investigation of opportunistic, range-limited communication between drones (leading to an imperfectly shared memory) and their effects on performance (measured by the time taken to discover and report a wildfire event).

We then move on to presenting another set of rules to be invoked in the monitoring phase, when the objective of the swarm switches from discovering a new event to establishing a surveillance perimeter and keeping the entire fire front under observation. This is achieved through a self-organization process that relies exclusively on local communication between, and decision-making by, individual units, and results in the emergence of a single continuous ring of evenly spaced drones through progressive rewiring of 'leader-follower' relationships. The proposed distributed algorithm is robust to the addition or removal of new swarm members, as well as to dynamic changes in the perimeter geometry (e.g., in the shape and size of an expanding fire). The presented work builds on the work done in the thesis by the lead author [1]. This work differs by exploring how the limited communication range between agents affects the efficiency of the swarm investigation. A new recruitment algorithm is also introduced to manage the division of labour to monitoring tasks of the mission.

1.2 Use-Case: Wildfire Detection and Monitoring

The work presented in this chapter is meant to be broadly applicable to a range of application domains but was undertaken in the context of a use-case specific domain: wildfire detection [2] and monitoring [3]. This research is funded by Academy of Finland (project grant number 348010) as part of the European Research Recovery Fund and conducted as part of the Unmanned Aerial Systems based solutions for real-time management of wildfires (FireMan) project. In this section we introduce the use-case.

Wildfires are devastating natural disasters that damage ecosystems and endanger human lives and property [4–7]. Their frequency increases with the intensification of climate change, and large fires can exacerbate climate change effects [8]. Therefore, it is important to limit the effects that wildfires have on nature and society as well as to develop systems for countering these effects to reduce the destruction they cause. The literature increasingly considers using unmanned platforms (e.g., [3,9]) to bear the brunt of the workload associated with regularly checking large swaths of forest for potential fires. Specifically UAVs are considered for this task [10], given that they can move fast and with little effort, even across territories that would be prohibitively difficult to traverse on the ground. In addition, airborne platforms have the advantage of the top-down view from which they can collect comparable data (such as video images taken at the same resolution and elevation above a location).

Firefighters' situational awareness may be limited by the remoteness or vastness of fire zones. Currently, the detection and monitoring of wildfires rely on various remote sensors deployed on multiple platforms, including drones [9,11–14], planes [7,15], and satellites [16–19]. However, all these platforms require highly trained crews or operators dedicated to their operation, and there may be limitations on how frequently information about the fire can be updated. Automating these systems could significantly increase their effectiveness in wildfire detection and monitoring.

Autonomous drone swarm systems [9] provide an effective solution for optimizing personnel resources within firefighting teams while improving situational awareness and fire monitoring. These drone systems can operate entirely independently, covering vast areas and continuously tracking wildfire progression without human intervention. The decentralized structure and use of relatively simple drones offer high redundancy in unpredictable environments, delivering robustness at a relatively low cost [9,10].

The use-case application presented can be seen as an (at least) two phase mission for a collective of unmanned platforms: initially the swarm is tasked with surveillance of an area with the goal of detecting any potential wildfire incidents as soon as possible. But once an event is detected the mission changes (at least for some members of the swarm) in that they are now no longer looking for other events but are instead switching to monitoring the detected event.

2 Materials and Method

This section details search and monitoring algorithms used in this chapter. It outlines at which point of the envisioned swarm mission these algorithms are used and presents the simulation tools used to validate the system's performance.

2.1 Modelling a Mission

As mentioned in Sect. 1.2, we distinguish phases in the swarm's operation. The envisioned swarm mission consists of four key phases, which happen serially, i.e., strictly one after the other, and are therefore discussed in isolation.

In this chapter we omit the obvious last phase where the swarm is recalled and returns to base. Furthermore, the actual *event* of detection a fire (a *Point*

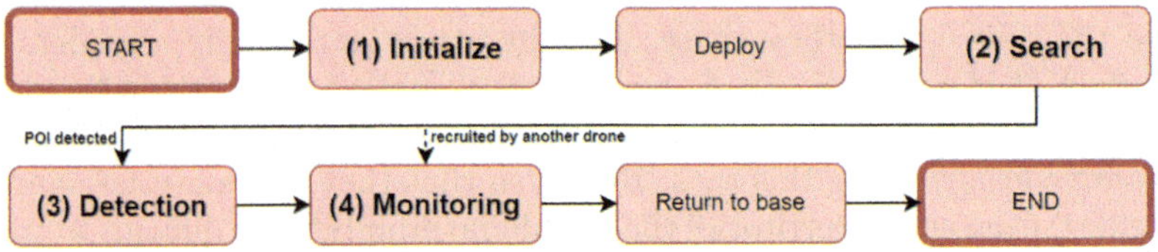

Fig. 1. The 4 key phases in the context of the whole mission: (1) `Initialisation` (2) `Search` (3) `Detection`, and (4) `Monitoring`; explained in this section

Of Interest (POI)) is a transition between phases (from `Search` to `Detection`): the `detection`-*phase* focuses on recruitment, i.e., the sharing of the information obtained by detecting the POI. We treat *'detecting a fire'* as an event (taking no time at all), but want to remain open to extending this to a complex operation. This would be driven by a need from the actual application (cf. [20] for a scenario where detection and identification of a target are achieved through the interaction of multiple, heterogeneous, platforms).

This chapter specifically addresses the algorithmic solutions for the swarming behaviour during the search and monitoring phases of the mission as these are the two phases where platforms achieve a common goal through interaction and collaboration. As we investigate the deployment of drone *swarms* for wildfire detection and monitoring, we focus primarily on the mission phases where multiple platforms collaborate. This includes spreading out to detect an event (`Search`) and subsequently cooperating to collectively `Monitor` the detected fire.

Phase 1 (`Initialisation`): Each unmanned platform undergoes an initialisation process where its systems are booted up, hardware is calibrated, handshakes are made and the drone announces itself to the dispatcher as being available for service. In turn, the platform is provided with information about other swarm members and the credentials needed to access services and infrastructure (note that the practical aspects of drone operation are omitted hereafter, as the chapter focuses on the higher-level swarming behaviour). Drones are also assigned a base as their starting location as well as a return point for recharging (also omitted in this chapter) and as their *home* to which they return after the mission. This 1st phase takes care of all practical issues (which are not the focus of this chapter), setting the stage for deployment (cf. Fig. 1) and the `Search`-phase.

Phase 2 (`Search`): After a drone is launched from the base it will engage in a specific, predefined behaviour (meaning the drone will execute a defined behaviour, chosen *a priori* by an operator and thus *not* dynamically determined by a mechanism within the drone). Both, the `Search`-behaviour as well as the `Monitoring`-behaviour are presented in Sect. 2.5. The goal of the former is to locate a POI, in the use-case chosen this POI is the source of a wildfire. To achieve this goal, a drone is inspecting locations from the map where any location that has not yet been expected has a chance of such a POI.

Individual drone behaviours are identical (cf. Sect. 2.5), and as a result (and by virtue of the fact that each drone in the swarm (a) executes the same behaviour and (b) because there is some form of interaction between the drones, causing an individual drone to react in their behaviour to the actions of others) the swarm, as a whole autonomously and efficiently divides the search area among the drones. This *division of labour* [21,22] has the potential for significant improvement in the overall performance (measured in the time it takes to locate a POI) while doing so with minimal communication requirements.

Notably, each drone operates on the basis of its own, subjective memory and we do not, *per se*, assume that information is shared between platforms. We do, however, allow for platforms to share the information about the map that they have gathered with others on the basis of proximity of the devices sharing the information, captured in the communication range R, cf. Sect. 2.4.

Phase 3 (`Detection`): There is a growing body of literature (e.g., [1,3,9,10]) on the automated detection of wild- or forest-fires. In this chapter we do not concern ourselves with the specific technique used to do so (cf. [2] for a recent publication discussing some of suitable approaches). For our work, the process of detecting a fire is reduced to a singular event that occurs when a drone reaches a location with a POI, and when it does, then the drone detecting the fire triggers the monitoring phase by notifying the swarm of the fire's location and recruiting additional drones to assist with surveillance. Internally, the drone then progresses in its mission from the `Search` phase to the `Monitoring` phase.

However, this phase is rather crucial for the swarm performance overall as it directly impacts (a) the number of drones that are progressing from the `Search` phase to the `Monitoring` phase and, consequently (b) the number of drones that remain to search for more POIs, making parameter R effectively a control parameter for the trade-off between *exploration* and *exploitation* [23,24].

When entering the `Detection` phase, a drone will broadcast the information that a fire was found to every drone within its communication range. Any drone receiving this message is thus automatically within a certain distance, defined through range R, of the POI. This ensures that recruitment is happening locally.

There are two limit cases: (1) when the range is zero, recruitment is non-existent (unless two drones share the same location), and (2) when the range is infinite (i.e., communication is possible across the entire map) then all drones will immediately enter the `Monitoring` phase and converge on the recently discovered POI. As a result (no drones remaining in the `Search` phase) no other POI can be found any more. Sections 3.2 and 4.2 provide the reader with results of investigating the trade-off between searching and monitoring, controlled by parameter R, as well as discussion of what that means, respectively.

Phase 4 (`Monitoring`): Without loss of generality, we require that any collective of platforms assigned to monitor the same POI is best arranged in a certain way around the POI to facilitate optimal data collection (i.e., monitoring). In our case we would like drones that cooperate in the monitoring of a specific POI to distribute themselves equidistant around the POI, with the rational being that this will provide the most diversity in the provided video feeds of the fire.

Furthermore, we want the swarm to remain in motion, effectively circling around the fire so that each drone itself is also continuously circling the event, the motivation here being that this way we ensure that all angles and views of the fire are captured by the sub-swarm. The process giving rise to this emergent property is described in Sect. 2.5, the results of investigating the impact in performance are provided in Sect. 3.4 and are discussed in Sect. 4.4.

2.2 Modelling the Map and Movement Through It

The map onto which the swarm is deployed is modelled as discrete locations, arranged in a hexagonal mesh (see Fig. 4 for a visual example). For any position (location) on that map, all immediately neighbouring locations in the mesh are considered to be connected to, and reachable from, that position (implying that there are at most 6 neighbours for any location, equidistant from it).

Movement on the map happens in discrete steps from one cell to an adjacent cell. Drone movement within the simulation is constrained to discrete transitions between neighbouring hexagonal cells, and the grid's geometry ensures uniform connectivity across the map. For the visualization and local memory tracking purpose the drone movements are tracked in Cartesian coordinates.

2.3 Modelling the Drones and Their Movement

The drone model used is identical for all drones. It consists of a single parameter R, defining the communication- (and this recruitment-) range of all drones and a local memory (for each drone) where it can record the locations it has visited, which serves as a list of all locations that have thus far *not* been visited (by that drone). Drones are furthermore assumed to collaborate whenever possible by sharing their memories whenever they are within range of one another. Doing so will extend the memory to contain the (subjective) memory of locations visited and currently unvisited, each time this happens a drone's memory can only (at worst) remain unchanged or (in any other case) improve the partial picture a drone has of locations already visited by the swarm.

The POI model determines the generation of wildfire incidents on the map, which are randomly assigned during map initialization, with all locations having an equal probability of containing a POI. These POIs are static and persistent over time, meaning they are allocated to specific locations at the start of the simulation and remain there throughout.

Both these constraints limit the applicability of our work to a range of applications. However, we argue that by considering static and persistent POIs we ensure that the results presented are meaningful in an abstract way as they are based on a minimal model with as little noise in the system as possible. Future work may include considering moving POIs (such as for example in Search-and-Rescue scenarios) or impose a half-life on incidents (such as e.g., when detection at an early stage is crucial for adequate response to be staged).

2.4 Communication

As mentioned above, all drones are assumed to have the same means for communication, which is circular and limited by range parameter R. The environment is idealized to be perfectly even and without any noise or interference, resulting in the area reached by a drone's communication to be a perfect circle. Visually, communication range is shown as circles around drones and lines connecting drones within range in the visualization shown in Fig. 4.

2.5 Modelling the Drone's Different Behaviours

Modelling the search behaviour starts by iterating between moving the drone and having the drone check for a fire (see Fig. 2). The latter is considered to be instantaneous for the sake of the simulations (but our work can easily be extended to model this process in more detail and complexity). The former is biased towards locations that have not previously been visited. A drone will maintain a record of the locations it has been to. Practically, this is a copy of the map (e.g., the one shown in Fig. 4) where all visited locations are marked (in Fig. 4 this is visualized by colouring them blue). In addition to this, drones may at times (whenever communication between platforms is possible, cf. Sect. 2.4) synchronize their maps to include locations visited by other drones.

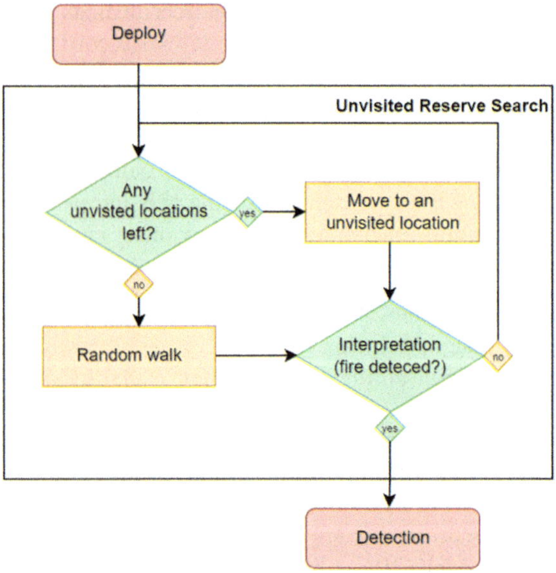

Fig. 2. Flowchart of the *Unvisited Reserve Search* algorithm. Round boxes, coloured red (Phase) are phases of the process, cornered boxes in orange (Action) are actions or processes and diamonds in green (Choice) are decision points.

A drone will identify the nearest location (measured in 'hops' over cells, not in Euclidean distance) that—as far as the drone is aware of—has not yet been visited, mark the location as *reserved* in its own memory (hence the name "Unvisited Reserve Search") and then move to that location. Upon arriving there the drone will first check whether the location is a POI (and if so, follow the protocol for recruitment, laid out in the next section). That failing it will mark the location as *visited* (in our visual representation this is in blue) and repeat the process by identifying the next nearest unvisited location. Once a drone runs out of unvisited locations it will simply perform a random walk.

Modelling the recruitment behaviour requires two variations: one for the drone recruiting and one for a drone being recruited.

If a drone detects a POI, it will transition from the `Search` phase to the `Detection` phase where it will communicate the location of the discovered fire in a broadcast to all other drones within range.

If a drone is still in the `Search` phase and receives an alert from a drone within its range informing it of a newly discovered POI, then the receiving drone will also transition out from the `Search` phase but it will directly move into the `Monitoring` phase (see Fig. 1). The motivation for recruitment (remember that without it, drones would continue their search indefinitely until they happen to stumble upon a fire) is that, for practical purposes, more than one drone is required to properly investigate a POI (i.e., the exploitation following the exploration).

Modelling the monitoring behaviour is reduced to arranging the drones allocated to a specific POI in a somewhat equally spaced arrangement around the POI. To this end, the drones (1) have to be arranged in an ordering where each drone has exactly one drone ahead of it and one behind; and (2) have to somehow equalize the distances between each other over time [1].

In the monitoring algorithm, each drone establishes a leader-follower relationship for themselves. This relationship is defined purely by the Euclidean distance. When a new drone joins the contour, the closest drone is designated as its leader. The leader drone then assigns the new agent as its follower. Subsequently, the new agent designates the previous follower of the leader drone as its own follower, while the old follower updates its leader to be the new agent. This can be seen in the visualization of Fig. 3.

The even distribution of drones around the contour is maintained by each drone calculating its distance to both its leader and its follower using Euclidean distances. If the leader is closer than the follower, the drone slows down slightly. Using Euclidean distance to manage spacing between agents removes the need for prior knowledge of the monitoring polygon's length or shape.

It should be noted that there is the unlikely case where a drone is recruited to a POI but on the way there encounters a new POI itself. If that happens, the drone abandons the earlier POI and instead attempts to recruit other drones to its POI, but, while doing so it can only recruit drones currently in the `Search` phase. This means that when a POI is discovered, there is (1) always at least one drone monitoring it (2) a potentially larger group of drones allocated to monitoring, and (3) a chance that the allocated group decreases (in the unlikely event described above) or increases (if another drone happens to stumble upon the POI) slightly over time. I.e., the sub-swarm is not fixed.

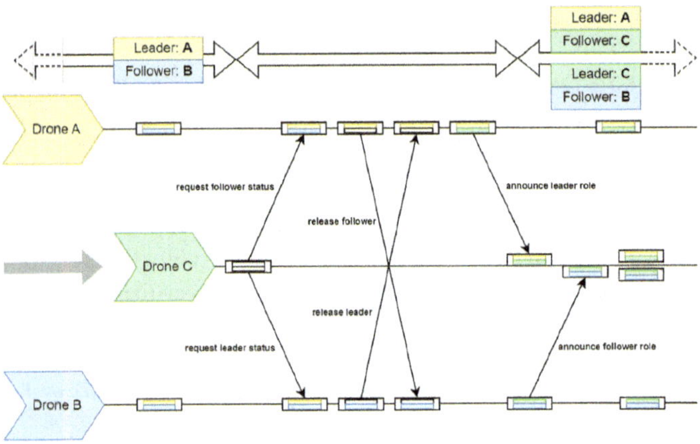

Fig. 3. Visualization of the process followed by a drone (here drone C) inserting itself into a contour to monitor a POI. This process only ensures a continuous contour, not equidistant positioning of the drones on the contour

To account for noise in the real world, an abstract polygon is created as a path for the drones to follow around the POI. All drones initially move to the closest point on the polygon and then start their swarm monitoring. In the real world, a drone would simply move towards its designated leader (by default the closest drone), thereby tightening the circle automatically, and then perform the monitoring tasks which would, have for different positions around the POI, a different ideal distance to it that offers optimal monitoring performance.

Furthermore, the goal is to have the swarm continuously move around the POI to continue to explore new angles and vantage points during the monitoring phase. This does imply forward movement as well as a direction, and as a default, our drones will always travel around a POI clockwise.

As stated above, a polygon is generated for each simulation run. The polygon generation algorithm (cf. Sect. 2.7) creates a random shape with a specified perimeter by iteratively adjusting vertex positions. It starts by placing vertices around a centre point using randomized angle steps (irregularity) and varying distances from the centre (spikiness). The perimeter is calculated by summing the distances between consecutive vertices.

2.6 Visualization

Each drone's movement and the visited cells are tracked visually in Fig. 4. The hexagonal grid is marked if a drone has visited that location. Drones are shown as red points, the POIs are marked in yellow, the green circle indicates the communication range, and the blue line shows if the drones are within communication range of each other.

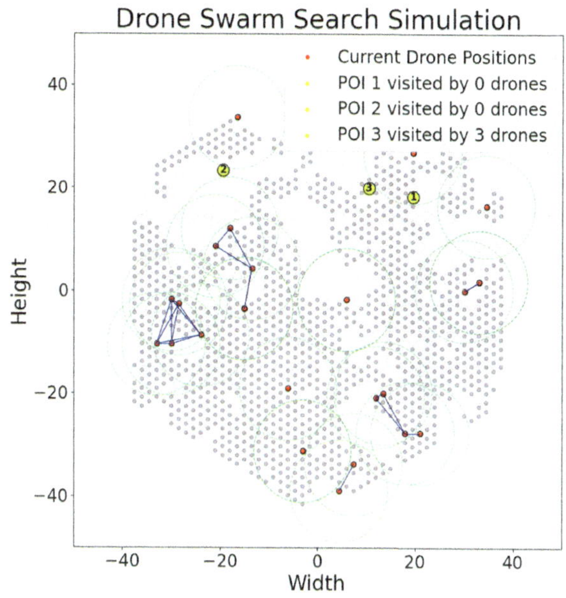

Fig. 4. A snapshot of the drone swarm simulation. Red dots represent drones, yellow dots are POIs, green circles show the communication range and blue lines indicate if the drones are within the communication range of each other

2.7 Numerical Experiments

Performance measures Two performance measures were used for the evaluation of the approach.

1. To assess the performance of the `Search` phase (where the swarm collectively explores the environment to detect POIs), we measured the relative probability of finding a POI. This is recorded as the percentage of all simulations in which POI was detected over time. This measure is in Sect. 3.1.
2. We also devised a scoring method that would allow us to keep a running score (overall performance value to date) to investigate how the swarm performance evolves as the swarm members switch from `Search` to `Monitoring`. The math is simple: we start with a score of zero, and then, at any time step, we (1) add 1 point to the score for every POI that has 3 or more drones allocated to it, and (2) we deduct 1 point for any yet undiscovered POI.

Swarm Size Versus `Search`-Performance of the Swarm focuses on the size of the swarm with regard to the detection of a single POI. The communication range R plays a big role and the investigations are repeated for different values for R to facilitate a discussion about its impact. To this end, we first performed experiments on a fixed map size (30 × 30 locations), running parallel simulations with two values for R ($R \in \{4, 16\}$). This was done for swarm sizes ranging from

2 to 24 in increments of 2, with 1000 independent simulations performed for each combination of settings. For each, we recorded the percentage of all simulations where that POI was detected and when. The results are plotted in Fig. 5.

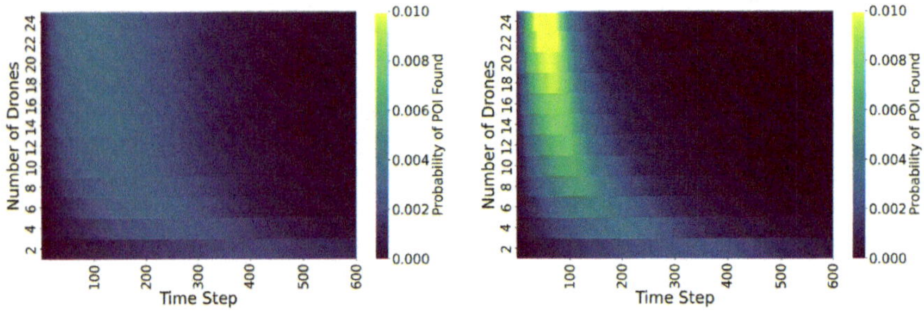

(a) With range 4 (b) With range 16

Fig. 5. The probability of finding the single POI. More drones improve detection, larger range increases the probability of detection

We then carried out a second series of experiments in which the size of the area varied (a square area of $n \times n$ with $n \in \{20, 30, \ldots, 100\}$). We ran 300 simulations for each area and for swarm sizes between 1 and 45 (increasing in increments of 1) and from 45 upward in increments of 5, and recorded the lowest swarm size that detected POI (in at least 90% of the 300 simulations). This was performed 4 times with $R \in \{4, 8, 12, 16\}$; Fig. 6 shows the results.

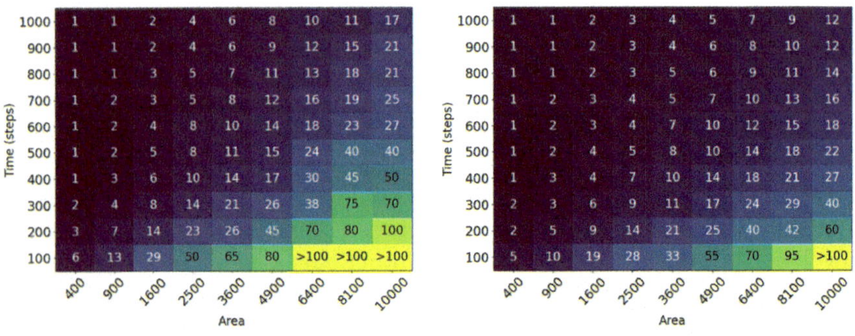

(a) Range 4 (b) Range 16

Fig. 6. The results of exploring, for $R \in \{4, 16\}$, the minimum swarm size required to detect a POI in at least 90% of all runs (300 simulations per cell)

Recruitment range versus `Monitoring`-performance of the swarm investigates the impact of changing R values for recruiting other drones to assist with the `Monitoring` phase. For these investigations a fixed map size (100 × 100 locations) was chosen, over which 20 drones and 5 POIs were randomly distributed. We recorded, for varying values for R ($R \in \{4, 8, 12, 16, 48, 96, 144\}$), when the last drone switched from the `Search` to the `Monitoring` phase (at which point the simulation ends). We also recorded the frequency with which any POI was undetected at the end of a simulation as well as the frequency of being monitored by 3 or more drones (the optimal number as it allows encirclement with the fewest number of drones). The corresponding results are plotted in Fig. 8.

Recruitment range versus the overall performance of the swarm Recruitment does, however, serve a practical purpose: many applications may emphasize the timely discovery *and monitoring* of events. To assess this aspect of the swarm performance we devised a cumulative scoring system that captures the swarm performance as a function of time, continuously subtracting points for undetected (missed) POIs, and adding points for POIs that are monitored by at least three drones. We vary the range, R ($R \in \{4, 8, 12, 16, 48, 96, 144\}$), using fixed values for drones (20), POIs (5) and a square 100 × 100 search area. The resulting performance curves are shown in Fig. 9.

Monitoring performance makes use of the scoring system discussed above to measure performance over time, as well as generates a polygon (see Sect. 2.5) to represent the noise path the drones follow when circling a POI to monitor it.

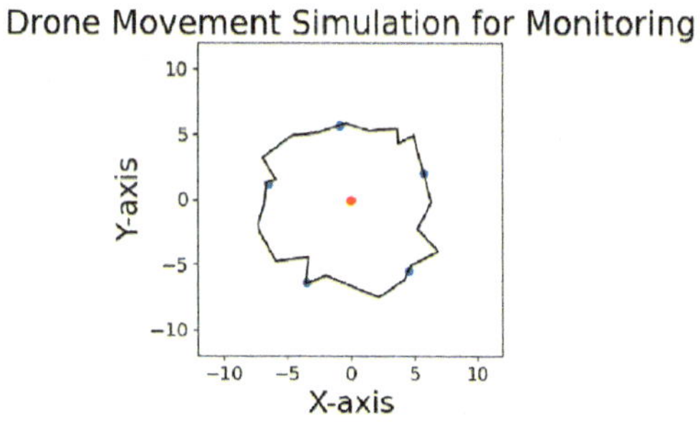

Fig. 7. An example of a randomly generated polygon around POI with 5 drones allocated to the POI. All polygons have a perimeter length of 50 and 25 corners

Figure 7 shows an example monitoring polygon with a total length of 50. While the length of polygons is a variable parameter, its contours always had 25 corners. The drones are initially inserted randomly along the monitoring contour, and the system will self-correct to ensure that the drones maintain equal spacing among the agents. Depending on the shape of the polygon, the distance along the contour will fluctuate around the mean of the normalized distance, which is $\frac{1}{n}$, where n is the number of drones. As we will show in Sect. 3.4, the more regular the polygon is the smaller the fluctuation around the mean is.

3 Numerical Results

3.1 Swarm Size Versus Search-Performance of the Swarm

The relationship between the number of drones, the number of steps, and the probability of finding a single POI is investigated, with Fig. 5a and b visualizing the relationship between the number of drones, the number of steps, and the probability of finding a single POI.

The communication ranges are 4 and 16 while the map size was fixed at 900 (30×30) locations. The x-axis represents the time progression in simulation steps, the y-axis corresponds to the number of drones, and the colouring shows the probability of finding the POI. It can be observed that when the size of the swarm is increased the time at which the POI is found decreases As the communication range increases the probability of finding the POI earlier increases. The relationship between how fast the POI is found and the swarm size is non-linear.

The impact of map size on finding the POI is further explored in Fig. 6: two values for R are considered as the area values are increased (x-axis). We record the time taken to find a POI. Reported are the smallest swarm sizes required to find the POI in at least 90% of the 300 simulations for each combination of parameters. As the area increases and the time required to find the POI within 90% of the simulations decreases, the swarm size must grow. Similarly, as the communication range increases, the time required to find the POI decreases.

3.2 Recruitment Range Versus Monitoring-Performance of the Swarm

The communication and recruitment range R effectively controls a trade-off between quickly allocating multiple drones to monitor a POI (for recruitment over large distances) and, on the other hand, prioritizing the search for additional POIs (recruiting very conservatively by only attracting drones over a small distance). The trade-off can become an *either/or* decision for the extreme values of R with zero communication effectively preventing any recruitment and a communication range covering the entire map enabling the first drone to find a POI to recruit every single member of the swarm immediately.

The values in-between have been numerically investigated through a number of simulations. The results shown in Fig. 8a confirm the intuitive expectation that the longer the recruitment range, the faster all drones switch from search to monitoring mode (which terminates the simulation since no more state changes are possible once no units are left patrolling the search area).

However, these results do not offer any insight into the end state in terms of how the swarm is distributed between the multiple POIs or of how many of these were missed entirely as a result of drones joining the monitoring effort early (at the expense of exploration). This is addressed by the results in Fig. 8b.

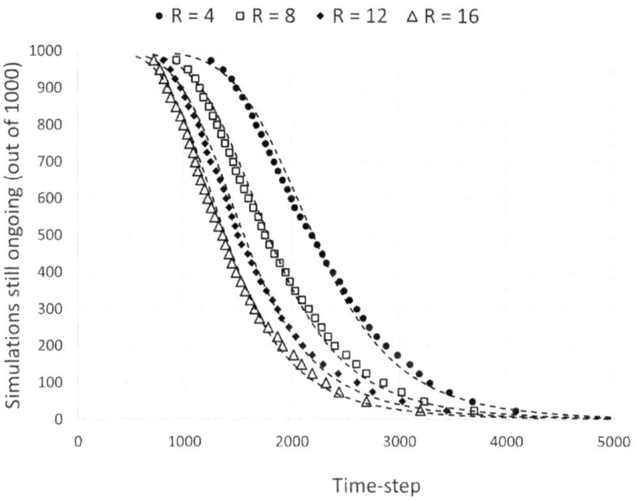

(a) Runtime of simulations (survival curve representation). Dashed lines are fittings using the Hill function (all $r^2 \geq .98$, indicating good statistical predictability by empirical means)

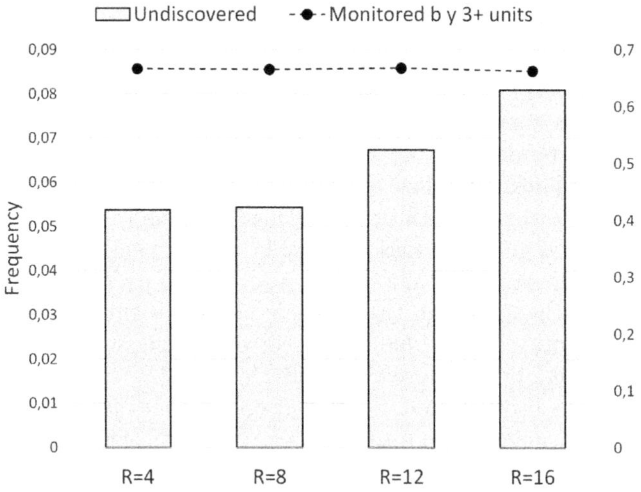

(b) Summary of the outcome of search-and-monitor mission.

Fig. 8. Investigating the impact of varying the range, R, for fixed parameters: 20 drones, 5 points of interests (POI) and a square 100×100 search area. The plotted results are the averages from 1000 independent realizations

3.3 Recruitment Range Versus the Overall Performance of the Swarm

On the other hand, the purpose of recruitment is not just to attract units to a newly discovered POI, but to do so quickly, the assumption being that the longer (or the earlier) a target is kept under observation, the better (in the case of containing a wildfire for instance, there is a clear benefit to establishing a cordon as quickly as possible following the discovery of a new point of ignition). Accordingly, we propose a composite score (performance measure) that also takes into account the time element and which works as follows: we start with a score of zero, and then, at any time step, we (1) add 1 point to the score for every POI that has 3 or more drones allocated to it, and (2) we deduct 1 point for any yet undiscovered POI. Any POI that has been discovered but is monitored by fewer than 3 units is considered *neutral* (no effect on the score).

The threshold of 3 was chosen in consideration of the parameter values, as it corresponds to one fewer drone than the perfectly balanced solution in the default scenario (4 out of 20 units orbiting each of the 5 POIs) and because a lower number prevents the sub-swarm from circling around the POI.

As illustrated in Fig. 9, this performance measurement method emphasizes the advantages and disadvantages of a longer recruitment range. For low values of R, the swarm struggles to quickly mobilize a sufficient number of units to monitor the POIs discovered in the early stages of the simulation, so the loss incurred by the still undiscovered POIs isn't compensated and the score drops to low negative values. In contrast, for high values of R, targets that are discovered early on (by random chance) are quickly *"taken care of"*, because strong recruitment allows the discoverer to quickly mobilize enough peers (at least two) to reach the target size for the monitoring team (3 drones). However, due to the stochastic nature of the process (how many drones are within recruitment range at the time of discovery is highly random), this target size can be exceeded (overprovision), which has a negative effect on the swarm's ability to subsequently detect and start monitoring other POIs (due to the corresponding reduction in the search team size), resulting in the score eventually reversing.

The overall conclusion is that, depending on whether the emphasis is on short- or long-term performance, opposite strategies (in terms of the recruitment range) offer the best result. Note: it would be relatively straightforward for a monitoring team to detect cases of overprovision and *release* members accordingly, but this function has not been implemented yet and will be the subject of future work.

3.4 Monitoring Performance

In Fig. 7, an example monitoring polygon is shown, Fig. 10 refers to this specific polygon and plots each drone's distance to its leader over time for one simulation run. The total length of the contour in this case is 50.

The drones are initially inserted randomly along the monitoring contour, and the proposed approach will self-correct to ensure that the drones maintain equal spacing among the agents. Depending on the shape of the monitoring contour,

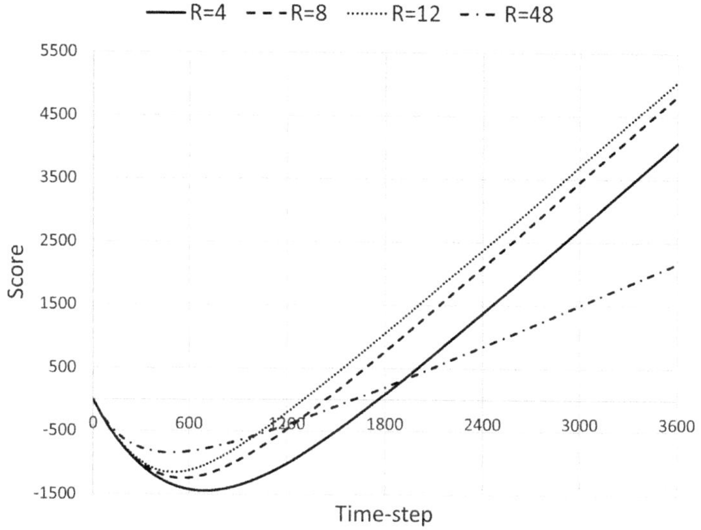

(a) The evolution of the overall performance score of a swarm.

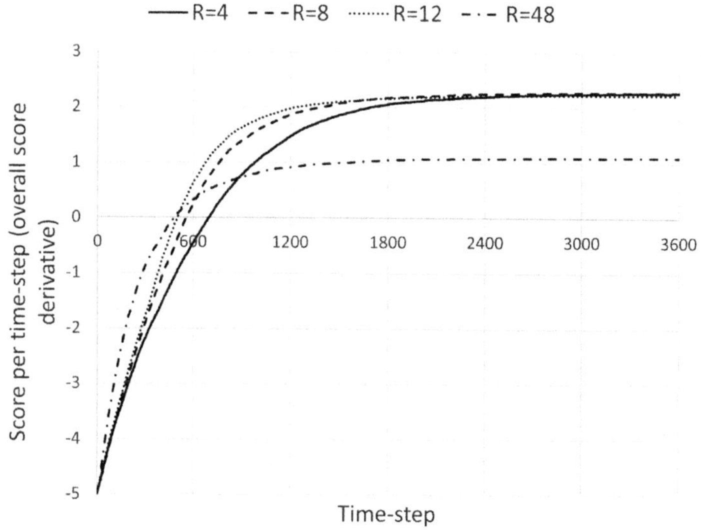

(b) Shown is the first derivative of the overall score.

Fig. 9. The evolution of the swarm performance over time, for varying the range, R, and the fixed parameters: 20 drones, 5 points of interests (POI) and a square 100×100 search area. The plotted results are the averages from 1000 independent realizations, a simple scoring system is used

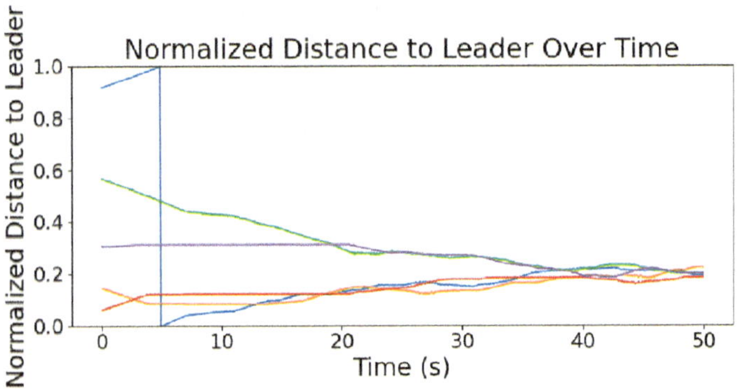

Fig. 10. Distance between drones during `Monitoring` (single experiment)

the distance along the contour will fluctuate around the mean of the normalized distance along the contour, which is $\frac{1}{n}$, where n is the number of drones. The more regular the polygon is the smaller the fluctuation around the mean is.

In Fig. 11 the normalized distance along the monitoring contour to the leader drone is shown for different contour lengths. In the plot, the solid line indicates the mean distance to the leader drone across all simulations, and the shaded area represents the standard deviation.

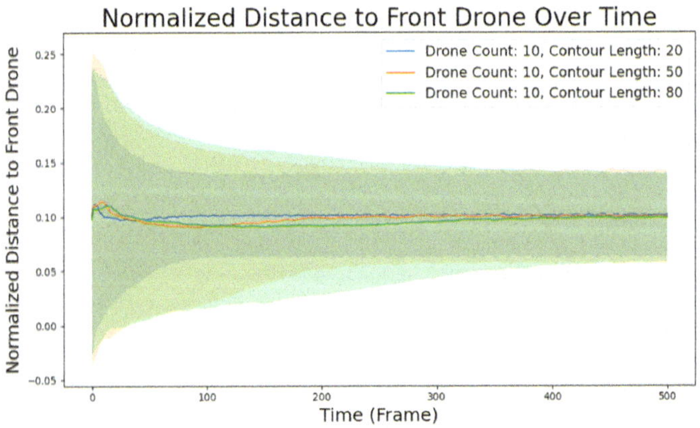

Fig. 11. Distance between drones during `Monitoring` (cf. Fig. 10): visualized is the normalized distance to the leader drone along the monitoring contour, for different contour lengths, across 500 simulations

A new random polygon is generated for each simulation. The distances from the drones to the leading agent are calculated and normalized. All distances to the leader drone have been averaged, and their standard deviation has been calculated; time is shown in arbitrary units. We observe the similar standard deviation in the normalized distance across the different contour lengths. However, with a shorter contour length, the system stabilizes more quickly.

4 Discussion and Conclusion

4.1 Swarm Size Versus `Search`-Performance of the Swarm

Figures 5 and 6 show the relationship between swarm size and search performance. Increasing the number of drones reduces the time required to detect a POI. This relationship is non-linear and the drones communication range plays a critical role in the efficiency of the search as larger communication range allows for better coordination among drones, leading to faster detection rates.

4.2 Recruitment Range Versus `Monitoring`-Performance of the Swarm

With regard to the impact of the communication range used for recruitment, Fig. 8a confirms the intuitive expectation that the larger the recruitment range, the faster all drones switch from search to monitoring mode.

However, the results shown in the two plots in Fig. 8 do not offer any information about the end state, i.e., how evenly the swarm is distributed over the detected POIs, nor do they capture how many POIs were missed entirely. These aspect are assessed by the results shown in Fig. 9, discussed next.

4.3 Recruitment Range Versus the Overall Performance of the Swarm

As illustrated by the plots in Fig. 9a, our cumulative scoring system amplifies the impact of a longer recruitment range. For low values of R (weak recruitment), the swarm struggles to quickly mobilize a sufficient number of units to monitor the POIs discovered early into the simulation, so the loss incurred by undiscovered POIs is not compensated and the score drops to low negative values. In contrast, for high values of R, targets that are discovered early on are monitored but, due to the stochastic nature of the process, this may have a negative effect on the swarm's exploratory performance due to overcommitment to the monitoring process, resulting in the score eventually reversing.

4.4 Monitoring Performance

Figure 8b provides some insight into this aspect but also doesn't paint a full picture. According to this representation, increasing recruitment range puts the swarm at a clear disadvantage, since the fraction of undiscovered POIs increases as a function of this parameter, whereas the fraction monitored by 3 or more units (which we chose as the success threshold since it is one fewer than a perfectly balanced distribution between all 5 POIs) remains near identical at ≈ 0.67.

The results shown in Figs. 10 and 11 offer insight into the stability of the monitoring algorithm: when the monitoring contour length is changed, the mean normalized distance to the leader drone remains the same.

4.5 Conclusion

This chapter explores the performance of self-organizing drone swarms that rely on limited local information sharing. It introduces a recruitment procedure to transition drones between the `Search` and `Monitoring` phases of a mission, while examining the balance between exploration and exploitation. The results indicate that increasing swarm size significantly reduces the time required for wildfire detection. The communication range plays a critical role in recruitment and monitoring efficiency. While a longer range speeds up the transition to `Monitoring`, it can also lead to overcommitting drones to a single POI, leaving others undetected. Achieving optimal swarm performance requires balancing `Search` and `Monitoring` tasks; our scoring metrics show the impact of recruitment range.

The proposed monitoring algorithm achieves a spatial distribution of drones around detected wildfires, allowing robust and adaptive surveillance of dynamic fire perimeters. Future work will focus on addressing limitations such as fixed POI locations and static communication conditions, as well as exploring dynamic recruitment strategies to mitigate overcommitment during monitoring phases.

References

1. Coffeng, S. (2024). *Decentralised drone swarm for wildfire detection and monitoring.* Master's thesis, Aalto University, School of Engineering, Espoo, Finland, September 2024.
2. Özel, B., Alam, M. S., & Khan, M. U. (2024). Review of modern forest fire detection techniques: Innovations in image processing and deep learning. *Information, 15*(9).
3. Tong, H., Yuan, J., Zhang, J., Wang, H., & Li, T. (2024) Real-time wildfire monitoring using low-altitude remote sensing imagery. *Remote Sensing, 16*(15).
4. Alizadeh, M. R., Abatzoglou, J. T., Luce, C. H., Adamowski, J. F., Farid, A., & Sadegh, M. (2021). Warming enabled upslope advance in western us forest fires. *Proceedings of the National Academy of Sciences of the United States of America, 118*(22), Article e2009717118.
5. Zhuang, Y., Fu, R., Santer, B. D., Dickinson, R. E., & Hall, A. (2021). Quantifying contributions of natural variability and anthropogenic forcings on increased fire weather risk over the western united states. *Proceedings of the National Academy of Sciences of the United States of America, 118*(45), Article e2111875118.
6. Williams, A. P., Livneh, B., McKinnon, K. A., Hansen, W. D., Mankin, J. S., Cook, B. I., Smerdon, J. E., Varuolo-Clarke, A. M., Bjarke, N. R., Juang, C. S., & Lettenmaier, D. P. (2022). Growing impact of wildfire on western us water supply. *Proceedings of the National Academy of Sciences of the United States of America, 119*(10), Article e2114069119.
7. Ifimov, G., Naprstek, T., Johnston, J. M., Arroyo-Mora, J. P., Leblanc, G., & Lee, M. D. (2021). Geocorrection of airborne mid-wave infrared imagery for mapping wildfires without GPS or IMU. *Sensors, 21*(9).
8. Carnicer, J., Alegria, A., Giannakopoulos, C., Di Giuseppe, F., Karali, A., Koutsias, N., Lionello, P., Parrington, M., & Vitolo, C. (2022). Global warming is shifting the relationships between fire weather and realized fire-induced co2 emissions in Europe. *Scientific Reports, 12*(1), 10365.

9. Yan, X., & Chen, R. (2024). Application strategy of unmanned aerial vehicle swarms in forest fire detection based on the fusion of particle swarm optimization and artificial bee colony algorithm. *Applied Sciences, 14*(11).
10. Saffre, F., Hildmann, H., Karvonen, H., & Lind, T. (2022) Monitoring and cordoning wildfires with an autonomous swarm of unmanned aerial vehicles. *Drones, 6*(10).
11. Thomas, P. J., Hersom, C., Al Kenany, S., & Staley, D. (1997) *Wildfire detection with a microsatellite* (pp. 633–640). Springer.
12. Moran, C. J., Hoff, V., Parsons, R. A., Queen, L. P., & Seielstad, C. A. (2022). Mapping fine-scale crown scorch in 3d with remotely piloted aircraft systems. *Fire, 5*(3).
13. Bushnaq, O. M., Chaaban, A., & Al-Naffouri, T. Y. (2021). The role of UAV-IoT networks in future wildfire detection. *IEEE Internet of Things Journal, 8*(23), 16984–16999.
14. Yuan, C., Liu, Z., & Zhang, Y. (2017). Aerial images-based forest fire detection for firefighting using optical remote sensing techniques and unmanned aerial vehicles. *Journal of Intelligent and Robotic Systems, 88*, 635–654.
15. Thomas, P. J. & Nixon, O. (1993). Near-infrared forest fire detection concept. *Applied Optics, 32*(27):5348–5355.
16. Toan, N. T., Cong, P. T. , Hung, N. Q. V. , & Jo, J. (2019). A deep learning approach for early wildfire detection from hyperspectral satellite images. In *2019 7th International Conference on Robot Intelligence Technology and Applications (RiTA)* (pp. 38–45).
17. Alkhatib, A. A. A. (2014). A review on forest fire detection techniques. *International Journal of Distributed Sensor Networks, 10*(3).
18. Dampage, U., Bandaranayake, L., Wanasinghe, R., Kottahachchi, K., & Jayasanka, B. (2022). Forest fire detection system using wireless sensor networks and machine learning. *Scientific Reports, 12*(1), 46.
19. Kelhä, V., Herland, E.-A., & Lohi, A. (2003). *Satellite based forest fire detection and automatic alert system—Pilot experiment* (pp. 649–653). Springer.
20. Saffre, F., Karvonen, H., & Hildmann, H. (2024). **(forthcoming)** the why, and how, of polymorphic artificial autonomous swarms. *Drones, X*(Y), 2024/2025.
21. Feng, Z., & Sun, Y. (2024). Research on division of labor decision and system stability of swarm robots based on mutual information. *Sensors, 24*(15).
22. Muhsen, D. K., Sadiq, A. T. V., & Raheem, F. A. (2024). A survey on swarm robotics for area coverage problem. *Algorithms, 17*(1).
23. Candelieri, A., Ponti, A., & Archetti, F. (2022). Explaining exploration—Exploitation in humans. *Big Data and Cognitive Computing, 6*(4).
24. Munir, A., & Parasuraman, R. (2023). Exploration—Exploitation tradeoff in the adaptive information sampling of unknown spatial fields with mobile robots. *Sensors, 23*(23).

Autonomous Drone Swarm Concept of Operations to Support Wildfire Management

Hannu Karvonen[1](✉), Eija Honkavaara[2], Juha Röning[3], Jukka Sassi[1], Jari Laarni[1], Vadim Kramar[4], and Fabrice Saffre[1]

[1] VTT Technical Research Centre of Finland, Espoo, Finland
Hannu.Karvonen@vtt.fi
[2] Finnish Geospatial Research Institute, Espoo, Finland
[3] University of Oulu, Oulu, Finland
[4] VTT Technical Research Centre of Finland, Oulu, Finland

Abstract. This paper provides an overview of a socio-technical Concept of Operations (ConOps), which can be used in future management of wildfires with the help of a semi-autonomous drone swarm. We developed the ConOps by using qualitative studies that included (1) seven interviews with subject matter experts, (2) a study of the literature, (3) a facilitated workshop led with human factors approaches, and (4) an expert feedback session. Firstly, the results of our paper summarize Finland's current wildfire management practices. Secondly, they highlight a few of the difficulties that drone usage in such situations faces today and potentially in the future. Thirdly, an overview of the main socio-technical elements of the developed ConOps is presented. The aim of this ConOps is to describe the basic setup of the operation of a drone swarm intended to support situation awareness and help in suppressing a wildfire by fire brigades. Fourthly, the results of a numerical experiment related to the fleet management of a drone swarm are given. Fifthly, a cost-benefit analysis on relevant human factors engineering activities in making this ConOps a reality is presented. Lastly, we discuss the scientific contributions of this study and suggest future research directions for further development and implementation of our ConOps into real-world operations.

Keywords: Drone swarms · Concept of operations · Human factors · Firefighting · Wildfires · Cost-benefit analysis

1 Introduction

The number, size, and intensity of wildfires are rising around the world, partly because of global warming. More effective wildfire-fighting solutions are needed to manage this growing threat. For instance, accurate and up-to-date geospatial data on the fire's location is essential to prepare the right firefighting tactics. To detect the fire and gather data on

Supplementary Information The online version contains supplementary material available at(https://doi.org/10.1007/978-3-032-19008-6_8) The videos can be accessed individually by clicking the DOI link in the accompanying figure caption or by scanning this link with the SN More Media App.

M. Laakkonen (ed.), *DroneAI for Resilient Nature*,
https://doi.org/10.1007/978-3-032-19008-6_8

it, the currently employed wildfire detection systems combine satellite and in situ technologies, such as aerial fire surveillance flights with small planes. Large wildfires may be successfully detected with satellite observations. However, small and newly started fires are usually not detectable via them due to the limitations of satellite imaging's geographical and temporal resolution. Although plane-based aerial fire surveillance flights may yield excellent situational information about smaller wildfires, the flights are typically labor-intensive, expensive, and rely on fossil fuels for power.

For real-time situational awareness of wildfires unmanned aircraft systems (UAS or drones) can be cost-efficient, environmentally friendly, and even disruptive tools that can significantly improve wildfire management in all its stages. The three main stages of wildfire management are prevention, surveillance, and extinguishing [1]. Firstly, the prevention phase aims at preventing fires and minimizing their effects [1]. The tasks of this phase may be planned, for instance, by using aerial remote imaging data that drones have obtained. Secondly, the fire surveillance phase has four main goals: (a) search of potential fires, (b) detection to alert firefighters, (c) diagnosis to get relevant data about the fire, and d) prognosis to predict fire propagation [2]. These activities are conducted to detect and monitor the wildfire as early as possible and mitigate its potential effects. All these goals may naturally benefit from the use of drones as "eyes in the skies" for situational awareness in real time. Thirdly, fire extinguishing includes suppressing efforts, such as building fire barriers, finding vehicle access and departure routes, locating adjacent runways and heliports, and extinguishing the flames with water [1]. Drones may be used also in this phase, even if they are not directly involved in the extinguishing operations. For instance, drone imaging can be used to find natural water supplies (to be used in ground-based extinguishing) close to the wildfire site or to gather information on local roads for instructions for the fire trucks on how to get quickly on site.

Given these examples, it is understandable why drones are currently being used in several countries to aid wildfire management efforts. Currently, for example, a dedicated firefighter may act as a pilot operating a single drone within visual range to capture aerial video footage of the fire. The pilot can share this footage with others, such as the fire brigade commander, to help coordinate the fire suppression efforts. In the future, this approach may evolve into the use of semi-autonomous drone swarms, operated beyond visual line of sight (BVLOS) from a far-located control center. This paper presents a Concept of Operations (ConOps) for such a semi-autonomous drone swarm to support wildfire management in future scenarios.

The results of this paper have been partially earlier reported in a conference paper [3]. The current paper extends this conference paper's results by: (I) providing more detailed results (e.g., Supplementary Online Materials [SOM] #1–#3 and Chap. 4.4); (II) providing new insights based on expert feedback about the first version of the ConOps in [3]; (III) providing results of a numerical experiment regarding fleet management performance of the drone swarm solution with the current features and prices of suitable drones (Chap. 4.5); (IV) conducting a cost-benefit analysis of relevant human factors engineering activities in making this ConOps a reality (Chap. 4.6); and (V) discussing the scientific contribution and further research directions (in Chap. 5).

2 Background

Several studies (see e.g., [4–6]) have already been previously conducted about the use of single drones to manage wildfires. However, the use of drone swarms for wildfire detection and management has received far less prior research. This lack of research is understandable, as the research, development, and application of non-military drone swarms in general are still in their infancy. In the next subsections, we will concentrate on some examples from the literature on the usage of drone swarms in wildfire management and then present some of the earlier Concepts of Operations work in the context of command and control of multiple drones in various domains.

2.1 Wildfire Management with Drone Swarms

Studies of wildfire management with drone swarms in earlier research can be divided into two distinct types: (1) fire detection or monitoring studies and (2) fire suppression studies. Here, we will focus on research on the former, as it is relevant to our study.

In [7], drone swarms for forest firefighting and the associated multi-level human-swarm interaction has been examined. [7] presents a case study focused on human factors work conducted in two Swedish fire departments that frequently use UAS in responding to fires. In this study, UAS operators and unit commanders designed a scenario that demonstrated how the swarm and its capabilities may be used based on their experience in using UAS in the context of combating forest fires and given the limitations and requirements of a forest firefighting mission. As a result, in [7] a model of swarm interaction was created to explain how the operators' engagement occurs at many levels, from the swarm as a whole to sub-swarms and individual Unmanned Aircraft (UA) to sensors and equipment that the UAs carry as payloads. The findings imply, for instance, that the related human-in-the-loop simulation studies should allow interaction between different swarm levels since this interaction can put the human operator(s) of the swarm(s) under more cognitive strain [7].

In [8], a study has been published regarding the viability and potential of using swarm robotics to fight fires with an emphasis on the self-coordination mechanisms necessary for the desired firefighting behavior to arise. The study created a self-organization system for swarms of firefighting drones with cooperative behavior based on a particle swarm algorithm and a physics-based model of fire spread. In order to demonstrate the effectiveness, scalability, and fault tolerance of the suggested self-organizing system, numerical tests were also carried out [8]. Similar findings on autonomous fleet or swarm control algorithms regarding fire detection and monitoring without human involvement have also been provided elsewhere, such as in [9–13].

Additionally, some earlier studies have suggested UAS swarm-based emergency management strategies. For example, in [14] a framework for a region's maximum area coverage during a wildfire event has been created. To improve the disaster response, [14] used an Artificial Bee Colony algorithm on the drone swarm to take pictures and collect data. The gathered photos were used to create maps of the burned area, identify entry locations, assess damage, and stop the fire from spreading. The goal of the suggested method was to optimize responses for exploration, exploitation, and to estimate the drones' maximum height for wildfire coverage [14]. The paper's authors also suggested a framework for Australia's State Emergency Services' emergency response [14].

Another relevant human control approach has been published in [15], in which the arrangement of trajectories for a fleet of fixed-wing drones to observe a wildfire evolving over time is presented. In order to give firefighters real-time information on the fire perimeters and their changes over time, [15] demonstrates a Situation Assessment and Observation Planning (SAOP) system that used a fleet of unmanned aircraft to monitor wildfires. The SAOP system operated along a Perception—Decision—Action scheme where the fire map and forecast were sent to the operators and subsequently used to define an observation plan that defines the optimal paths for the fleet to observe the fire. The fire map was updated using the recently acquired data, and the resulting trajectories were transmitted to the UA for execution. In the presented system, supervision was playing the role of a thin abstraction layer issuing high-level commands to components. Although this gave the users some operational autonomy by removing the need to define low-level tasks or trajectories, adding a higher planning layer would require further monitoring capabilities. Together with the higher-level mission planner, this method might, for instance, enable human operators to determine whether monitoring plans can distribute resources aggressively or conservatively and assist them in controlling the long-term fleet capacities [15].

Lastly, the results of a systematic numerical techniques (Monte Carlo simulation) analysis of drone swarm behavior for wildfire monitoring have been presented in [16]. This study shares findings about how key simulation parameters, such as swarm size, surface area being observed, and fire propagation dynamics, affect how well an autonomous wildfire drone force performs when left to operate without human oversight. Moreover, the study emphasizes the use of straightforward, reliable, and practically implementable distributed decision functions that can assist the drone swarm's self-organization in achieving a group objective. The study's findings supported the existence of significant non-linear effects in the interplay between the previously described key parameters, which may be closely approximated by using a specific empirical law [16].

2.2 Earlier Concepts of Operations Work for the Command and Control of Multiple Drones

The notion of Concept of Operations has been proposed, for example, in [17]. According to [17], a ConOps document includes a description of the current system or situation, justification for and nature of proposed changes and/or new features, operational concepts for the new or modified system, operational scenarios for that system, a summary of organizational and operational impacts, and an analysis of the proposed system. In earlier studies (e.g., in [18]), developing a ConOps has been demonstrated to be a beneficial and integrative element for various stakeholders in the design of complex socio-technical systems. For instance, a ConOps has been demonstrated to mediate interactions between different stakeholders by offering diverse illustrative viewpoints on the socio-technical system being designed and facilitating cross-disciplinary discussion among specialists from diverse backgrounds.

Examples of related domains where ConOps work has been carried out include the military domain (e.g., [18, 19]), civil aviation (e.g., [20]), as well as UAS traffic management (e.g., [21, 22]). There is also earlier research reporting ConOps specifically for multiple (semi-)autonomous drones. Next, we give a few examples of this literature.

In order to determine the informational and functional requirements that meet the demands of stakeholders dealing with UAS operators as well as their collaborative needs, [23] outlines the creation of a novel design requirements analysis method. This study expands on a previously established requirements analysis technique known as the Hybrid Cognitive Task Analysis method in their work. In the study, the method is extended by introducing analytic steps to identify task and decision-making dependencies between different UAS operations collaborators. The notion of boundary objects, a popular analytical construct in the study of group activity, is also employed by the authors. Boundary objects are physical or informational artifacts that transcend task boundaries between members of different groups. A ConOps can also be seen as a particular kind of boundary object. Identifying boundary objects in complex task operations helps the analyst, for example, to identify task and decision-making dependencies between local and remote collaborators. As a result, knowing these dependencies aids in deciding the information-sharing needs that the UAS ought to fulfill [23].

Furthermore, a ConOps for UAS in mapping the devastation caused by Hurricane Katrina has been defined in [24]. This research looks at small unmanned aircraft system (SUAS) vertical take-off and landing (VTOL) operations that were part of an 8-day structural inspection mission after the 2005 hurricane. Key findings for the developed ConOps are drawn from the observations. A crewing organization and flight operations strategy for SUAS are suggested considering the results and additional observations [24]. As a result, in addition to its technical components, the ConOps incorporates organizational and operational activity elements.

Lastly, the development of a ConOps for a swarm of autonomous robotic platforms (such as drones) in the military domain has been presented in [25]. The goal of this study was to show how autonomous robotic swarms could be used in many military branches in the future. When designing, validating, or purchasing an autonomous robotic swarm system, the suggested ConOps can be utilized as a boundary object. Furthermore, the ConOps should be maintained as an overview description and definition of overarching goals and policies throughout the system lifecycle. The ConOps approach in our study aligns well with this human factors-oriented approach in [25].

Despite the fact that there is undoubtedly some prior research on Concepts of Operations for several drones in various sectors where they have been effectively implemented for specific applications, a systematic approach to ConOps development appears to be lacking. Specifically, the examined literature omits information about the meticulous planning of the human operators' tasks, such as how they are planned to supervise and manage the drones while on particular missions.

2.3 Cost-Benefit Analysis as Part of ConOps Development Work

The value of a drone swarm platform can be defined as the worth of the proposed solution in relation to its purpose and operational context. The platform must provide value for money: the monetary costs and other resources need to be appropriate to achieve the desired outcomes [26]. Therefore, cost-benefit analysis (CBA) and life-cycle cost estimation can play an important role in a ConOps, even though they are quite often missing from these descriptions. The aim of a CBA is to demonstrate that the operational benefits outweigh the financial costs, and it involves a systematic comparison between the incremental costs and incremental benefits of the proposed solution. For example, benefits can be opportunities for savings through the implementation of the new solution, and they can be characterized either as value added or potential risks avoided. Examples of cost-benefit estimation of future air transportation and air traffic management concepts have been presented, for example, in [27–29].

More specifically, CBA can be used in the design stage for the evaluation of engineering activities needed to reduce the potential risks throughout the system lifecycle. Costs and benefits are typically actualized at different points in time, and they can include past, present, and future costs, and benefits [30]. Therefore, the implementation cost of a particular system should be considered from the system lifecycle perspective. For example, since the value of costs and benefits changes as time elapses, a discount rate is typically used in the CBA to adjust costs and benefits to their actual present value [31]. The lifecycle-cost estimation is also related to the management of risk in a system's lifecycle. As stated in [32], risks are at their largest immediately before/after the implementation of a new system but so are also the opportunities for benefits.

3 Study Case Approach

This study is part of the FireMan project (unmanned aerial systems based solutions for real-time management of wildfires). In the FireMan project, we use the ConOps document as a knowledge artifact that is created during the early stages of system development. Nevertheless, we see that it can be used also in all the phases of systems engineering and the lifecycle of a system. In practice, we view ConOps as a transitional design artifact that is essential, particularly during the requirements formulation phase. Additionally, we have used the ConOps as a template that can be modified and updated as the development proceeds regarding specific user needs and use cases [25]. As has been done in another similar project in the past (see [33]), the term ConOps is used in the context of our study case to refer to the design target that comprises the work organization, human activities and key tasks, and the related technology. It is, therefore, different from the Concepts of Operations that are developed, for example, in the military domain, which typically describe what a joint force commander intends to accomplish and how it will be done using available resources [19]. Additionally, our ConOps methodology differs from those required of all UAS operators as the initial step in Specific Operations Risk Assessment (SORA) procedures [34] and related activities that are utilized to acquire a UAS operation authorization. Our ConOps methodology is more similar to, for example, the U-space ConOps [21], which outlines how operations should take place in very low-level airspace specifically from the viewpoint of the user.

In order to build a future ConOps, a thorough understanding of current operational practices is required. We conducted seven subject-matter expert interviews (lasting 1.5–2 h each) and a large online full-day workshop with 40 participants from relevant Finnish stakeholders (such as fire departments, authorities, businesses, and research organizations) to gain an understanding of the current ConOps of wildfire (and particularly forest fire) management in Finland. The Core-Task Analysis [35] and other scientific human factors research methods served as the foundation for both the interviews and the workshop. In practice, this meant that the interview and workshop questions focused more on the human activity-related practices instead of technological factors.

The interview questions and workshop themes included parts related to, for example, (1) current aerial wildfire surveillance practices, (2) current main activities in a wildfire management situation in a given scenario, (3) current key actors/stakeholders and knowledge entities in a wildfire situation, (4) different challenges of drone usage in wildfire situations, and (5) future visions of drone usage in wildfire situations. The scientific background methodology used in the investigation of these themes with the participants varied, but a key approach used in mapping the current main activities in a wildfire management situation was Functional Situation Modeling (FSM).

The FSM technique has been employed in the past literature to analyze operating procedures in complex work (see, e.g., [36, 37]). The FSM method connects critical functions from the domain in question with the operational activities being performed. The approach is formative in nature and enables the investigation of operational activity from the perspective of behavior and work practice. The method aims to promote resilient work practices by identifying actions in specific situations that contribute to the overall goal of safety [36]. The resulting output denotes a control situation from the point of view of its critical functions, which are typically endangered in the specific safety-critical conditions. This output also shows the related human activity and links this activity to these critical functions that should be supported during the work activity. The operational activity in these types of environments is typically uncertain, dynamic, and complex, which requires keeping the operational focus on the core task at all conducted activities [35]. The basic schema of the FSM notation is shown in Fig. 1, which has been modified based on [37]. By applying this predefined notation technique, the participants in our workshop began working collaboratively by outlining the major objectives, functions, and phases of the fire rescue services in the defined scenario.

A ConOps for using a semi-autonomous drone swarm to support wildfire management was created based on the data acquired from the interviews and workshop. The developed ConOps describes what its main socio-technical elements are, who are the main stakeholders and actors, and how are the basic operational activities organized.

Identifying the relevant operational scenarios and defining the corresponding requirements for human-swarm interaction have also been crucial to our ConOps development work. The ConOps of the entire human-drone swarm socio-technical system that has been established includes easily comprehensible documentation, graphics, and animations that explain the features and intended use of the proposed swarm system from the perspective of its human users. We also assessed and specified the roles of human operators, their primary responsibilities, their interactions with the semi-autonomous drone swarm system, and their collaboration with relevant stakeholders from the perspective of human factors.

After developing the first version of the ConOps [3], we organized a feedback session on it where approximately 50 experts could express their views on the potential gaps and limitations. The final FireMan future ConOps presented in Chap. 4.4 of this paper was developed based on this feedback.

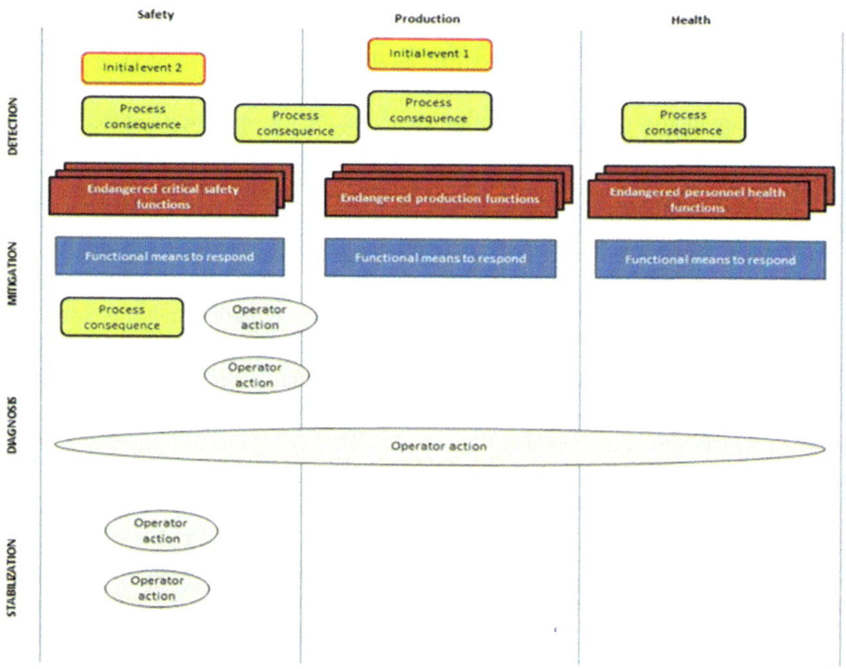

Fig. 1 A general schema of an FSM. In modeling a scenario, the boxes describe: process events and information (yellow), critical functions (red), and operator actions (white). The actions are connected to the critical functions (Safety, Production, Health [38]) and the operative means (blue) to mitigate the functions. The description is organized in chronological phases (detection, mitigation, diagnosing, and stabilization) according to the scenario in question [37]

To further answer the feedback received, we conducted a numerical experiment of the estimated number of required drones (considering the fleet management) of the swarm solution with the current features and prices of a suitable drone (see Chap. 4.5). Additionally, the human factors engineering CBA for this ConOps was conducted with a purpose-built tool, which has been previously utilized in the nuclear power domain [39]. The method for using this tool consists of five steps: First, risk influence factors that are included in the calculation are specified; Second, weights are determined for the risk influence factors according to the properties of the project; Third, the proportion of the budget allocated to engineering work is determined; Fourth, budget shares are calculated by multiplying the weights with the proportions of the project budget; Fifth, the budget share allocated to engineering work is calculated by adding up the percentages of budget elements. The results of this CBA are presented in detail in Chap. 4.6.

4 Results

Four main results entities emerged from the conducted interviews and workshop: (1) the current operational activity, (2) the possible challenges of drone usage in a wildfire situation, (3) participants' future visions of drone usage in wildfire situations, and (4) the final version of the developed drone swarm-based FireMan ConOps. Furthermore, based on the feedback on the first ConOps version, results of a (5) numerical experiment related to the fleet management of the drone swarm and of (6) a cost-benefit analysis on relevant human factors engineering activities in making the operational concept a reality are also presented. We have next divided these results into separate subsections.

4.1 Current Operational Activity

The basic operational activity in a wildfire situation includes two distinct stages: (1) wildfire detection and monitoring and (2) wildfire management (including suppression activities). The current detection activities in Finland are conducted with aerial wildfire surveillance flights around the country during summertime when the wildfire index goes above a certain threshold. These flights include the following phases (in an actual wildfire situation): (1) Flight route planning; (2) Take-off; (3) Flying the planned route; (4) Smoke detection; (5) Locating the source of the smoke (reconnaissance); (6) Alerting the emergency center (in case of a real wildfire); (7) Assessing the fire area; (8) Mapping the roads, natural water sources, and guiding the fire brigade; and (9) Producing a situational picture for the incident commander on the ground as the fire proceeds. The related main operational aims and additional relevant information for each of these phases are presented in detail in the Supplementary Online Material (SOM) #1 (see link).

Once a wildfire has been verified, several main actors (i.e., organizations responsible for their own part) become important in responding to the situation. In our subject-matter expert workshop, we mapped these stakeholders and their primary responsibilities during a wildfire scenario. The task descriptions of the different main actors and some relevant additional information are presented in SOM #2 (see link) of this paper.

Additionally, we identified both static and dynamic information entities that are pertinent in a wildfire scenario and what this information typically consists of (i.e., examples of its detailed data). SOM #3 (see link) provides a list of these different information entities, examples of the data they include, and their types (static/dynamic) in detail.

Following the definition of these basic information entities, the workshop participants were shown a hypothetical operational scenario (see the "Initial Situation" section of Fig. 2) and an initial Functional Situation Model schema (created by the workshop organizers based on the earlier interview results), which was applied to our case scenario. The final version of the FSM for the hypothetical scenario we had developed and walked through together with the experts in the workshop is presented in Fig. 2. In the FSM description, the main aims of human activity in this context are the protection of human lives, the environment, and property (the red boxes, see details in [37, 40]). The operational functions (the white boxes) of the fire rescue services during this type of wildfire scenario contribute to one or several of these main aims. However, a description of which functions contribute to which main aims is not included in this paper.

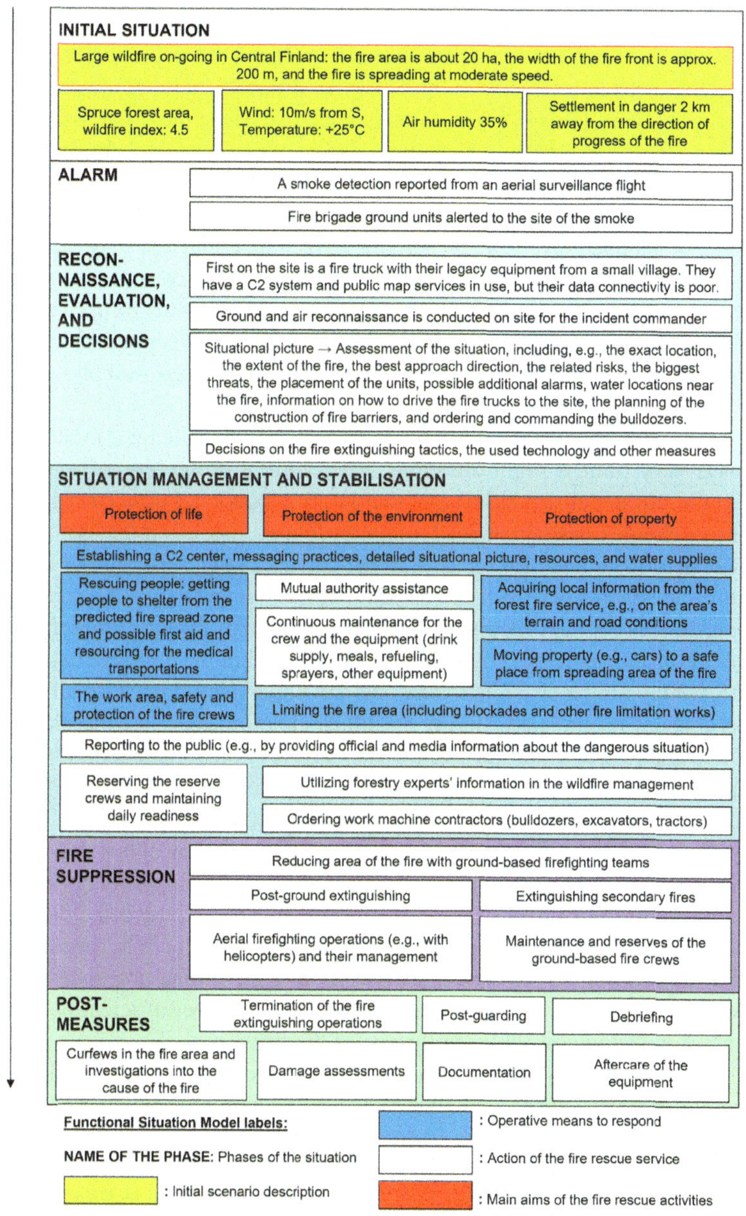

Fig. 2 An FSM of the main phases and measures taken in a specific chosen wildfire situation

4.2 Challenges of Drone Usage in a Wildfire Situation

A broad range of technological and operational challenges specifically for Nordic UAS operations have been studied and categorized earlier in [41, 42]. Our case study results revealed that there are a multitude of other factors that may currently cause challenges, such as unexpected situations, during drone flight missions in a wildfire scenario in

Finland. The identified main challenges are briefly presented next. These challenges should be overcome in order for the suggested ConOps in this paper to become a reality.

Firstly, there can be problems in operating the drone, because of the smoke and fire. In visual line-of-sight (VLOS) operations, smoke can cause the drone pilot to lose the visual contact with the drone, which may lead it to accidentally collide with nearby trees and plummet from the sky. Additionally, if the drone is flown too near the wildfire, the heat might damage the drone's sensors, navigation system, electronics, materials, or structures. The planned drone-based activities and data collection may be thwarted by the greasy smoke from the fire, which can also obstruct the optical lenses used for remote sensing. Consequently, the drone's situation awareness data cannot be used to coordinate the firefighting activities. Therefore, the drone should be flown at a safe distance from the flames of the wildfire, its flying height should be maintained well above the nearby trees, and it should not enter the airspace near the smoke.

Secondly, the current power solutions for UAS present challenges: modern electrical drones have very limited flight durations. Therefore, a drone reserve is necessary to ensure continuous aerial surveillance of the wildfire location. Furthermore, insufficient drone battery charge capacity can lead to shorter flight times, and if this is not considered in advance, the longer need for recharging the batteries may cause further delays.

Thirdly, the benefits of drone-based wildfire management may not be realized due to a lack of resources for the processing and interpretation of the airborne remote sensing data. Even if the data collected by the drones would aid in the effective extinguishment of the fire, it may not be possible to analyze and understand it because all the relevant personnel are busy in other tasks during a wildfire situation. To ensure that skilled experts are available and to prevent drone missions from relying on a single individual who may be occupied with other tasks during a wildfire, different types of drone missions and their gathered data analysis should be carefully planned beforehand.

Fourthly, drone flights may be delayed or even prevented by either inadvertent (e.g., due to a lack of coverage) or deliberate (due to jamming) disruptions in the GNSS (Global Navigation Satellite System) or data connectivity links. For the intended wildfire-fighting activities, the used positioning and communications systems should be robust, redundant, and sufficiently secure. To stream the data from the sensors and prevent the loss of the data link between the drone and the pilot, there should also be sufficient bandwidth and low-latency connectivity provided at the site of the fire.

Fifthly, as autonomy of drones advances in the future, especially surveillance-related operations with a high level of autonomy and low level of human involvement need to be carefully designed from the human factors' perspective to avoid the human operators' distractions or boredom during the operations: there should be enough meaningful tasks for the human operators to keep them alert and engaged in the operations.

Lastly, integrating an autonomous drone swarm solution with the systems used by the fire departments and aviation authorities would be one of the biggest hurdles to overcome. In order to have the drones and crewed aircraft flying in the same airspace, the integration calls for innovative UAS traffic management (UTM) technologies. Furthermore, for the authorities to receive and display the situation awareness data gathered from the drone swarm sensors and the related sensor fusion platform, a great deal of integration work is required with the application programming interfaces (APIs) of the rescue information systems and other solutions that the firefighters are using.

4.3 Visions and Possibilities of Drone Usage in Wildfire Situations

We established the following three distinct time frames for the workshop's brainstorming sessions about the use of drones in a typical wildfire scenario: (I) the use of drones in a wildfire situation at the moment, (II) the use of drones in the near future (1–5 years), and (III) the use of drones in the far future (5–10 years). Within each time frame, we thoroughly went through various relevant subjects, such as what kinds of drones can be used, who operates the drones, what is the level of automation, and what cameras/sensors can be used onboard the drones.

In summary, the following results on the participants' perceptions of drone use during wildfires over the selected time periods (I–III) was gathered. (I) In the current situation, to provide the fire department with video and thermal imaging camera footage of the fire situation from the air, light and inexpensive commercial quadcopters are flown at quite low altitude in VLOS mode, primarily by individual pilots. This gathered data is not currently analyzed with automated technological means (e.g., with artificial intelligence, AI), but only by humans with the aim of detecting the fire front and its spreading direction and speed. When directing firefighting teams in the field and their extinguishing efforts, this information can be used to coordinate the efforts. In general, the existing operating procedures lack real-time analysis of the wildfire situation. In Finland, the data transfer in the wildfire site is reliant on either public or private mobile networks. Examples of public networks include 4G mobile networks and the TETRA-based (Terrestrial Trunked Radio) VIRVE network. These networks may also have issues with coverage and bandwidth, which can result in a bottleneck effect for data communications. For basic airspace management, for instance, temporary R (restricted) or P (prohibited) airspace areas can be used. While certain automated solutions (e.g., for specifying the flying waypoints of the drone beforehand automatically) are now in use, autonomous systems are not yet used. One key challenge is the sharing of situation awareness information efficiently from the drones to all relevant stakeholders.

(II) In one to five years, the participants saw heavier (both fixed-wing and quadcopter) professional drones flying semi-autonomously in BVLOS mode and using their sophisticated sensors to provide a high-quality real-time situational picture from a high altitude. To make this scenario a reality, however, extensive drone operations training is required. Furthermore, the participants saw that satellite-based remote imaging or even tiny airships might be used for long-term aerial observation of the situation. Several participants also suggested that in a wildfire scenario even drones that suppress the fire may be utilized. Additionally, firefighters on the field could get vital supplies (such as tools or power banks) delivered by drones. On the situation analysis side, many stakeholders might be given access to a real-time map depiction of the present fire condition and its predicted propagation during certain time intervals. Advanced sensor fusion would also allow detailed analysis of the fire situation by combining different sensor information into an intuitive and coherent whole for the humans. As a result, firefighters may be effectively commanded and controlled to carry out suitable predictive extinguishing operations. The data connectivity to the wildfire site is expected to be provided through advanced and wide-coverage private, public mobile (e.g., VIRVE 2.0 and 5G), satellite communications, and Internet of Things networks. With the right UAS traffic management solutions, air corridors, and automatic detect and avoid technologies, safe airspace management could also be carried out even in populated areas.

(III) In five to ten years, wildfire detection and monitoring activities across wide geographic areas may be accomplished by (semi-)autonomous drone swarms. These swarms could be deployed automatically from drone base stations in 24/7 readiness spread around the country. With improved machine vision and AI capabilities, the swarms may automatically identify and reconnoiter wildfires and offer relevant stakeholders easily comprehensible real-time information about the fire. To provide relevant situational awareness information, enhanced sensor fusion might be paired with hyperspectral and other advanced imaging from various angles in addition to standard visible light RGB and thermal imaging. Additionally, heavy-duty drones may help fire teams with various fire extinguishing operations and logistics. The drones' command and control, along with the associated analytics, are centralized and semi-autonomous. Additionally, with the aid of AI, comprehensive forecasts of the fire scenario at certain future time points could be made. Secure and reliable 6G networks could be used to give data connectivity in the fire area. Advanced UTM systems can also be used to regulate the airspace, enabling both crewed and uncrewed aircraft to fly safely in the same area.

4.4 FireMan Drone Swarm-Based ConOps

To create the basic building blocks of the FireMan Concept of Operations, we conducted a co-design session with the participants in the workshop, as has been detailed in Sect. 3 of this paper and in [3]. Based on the results of this co-design session, we developed the first version of the ConOps (see [3]). The final version (in this paper) of the drone swarm-based FireMan ConOps was created based on the comments and results we obtained in a separate expert feedback session regarding the earlier version of the operational concept. The main feedback regarding the first version of the ConOps [3] was related to the costs of developing and acquiring such a system. Therefore, we conducted both an analysis of the number of needed drone units (Chap. 4.5) for the swarm for the described scenario (see Fig. 2) as well as a cost-benefit analysis regarding relevant human factors engineering activities in making this ConOps a reality (Chap. 4.6). Next, we provide an overview of the key elements of the developed final ConOps.

Process description. Figure 3 shows a high-level overview of the operational phases involved in detecting and suppressing a wildfire in the FireMan ConOps. This type of a process is expected to be possible in approx. Four to eight years in Finland (at Technology Readiness Level [TRL] 8) and is explained in textual format next with references to the phase numbers of Fig. 3. First, a fire source is discovered in the environment by a manned aircraft fire surveillance flight, a long-range drone, a drone swarm, or a person on the ground (1.). Second, a more thorough identification of the type of the fire is required (e.g., is it a true wildfire or is someone burning trashes in a controlled manner, (2.). At this stage, the aerial reconnaissance of the fire situation could be conducted using a drone swarm equipped with sophisticated detecting capabilities. Third, if it is a real wildfire, the drone swarm would begin tracking its propagation from various angles and help would be alerted (3.). Both fixed-wing and quadcopter drones flying in a swarm would be equipped with various remote sensing capabilities. The fire department crews would also start their necessary activities after arriving at the location via ground routes.

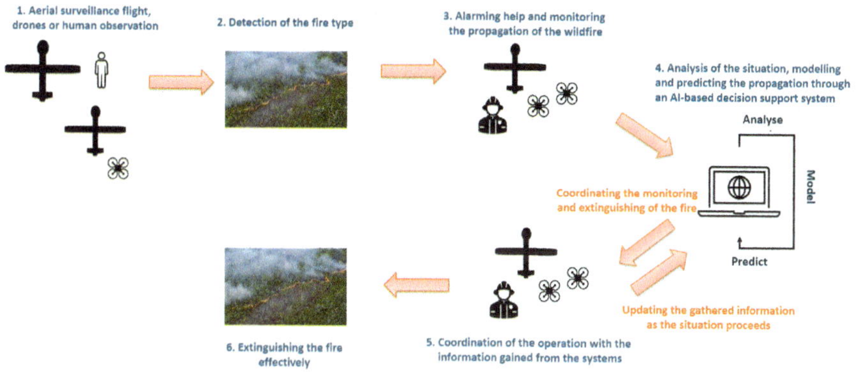

Fig. 3 High-level process description of the operational activities when detecting and extinguishing a wildfire in FireMan ConOps

Fourth, a centralized AI analysis system receives the data gathered from the drone swarm's sensors and uses it to create a digital twin that simulates the fire and its spreading (4). Real-time automated situational analysis and future fire behavior prediction would be made possible with this digital twin. In this manner, it would be possible to forecast how the fire would propagate over specific time periods (e.g., in five, 15, or 30 minutes) if no extinguishing measures were taken.

Fifth, an AI-based decision support system processes this data and can offer suggestions to the fire situation's incident commander to coordinate the firefighting teams and their extinguishing operations in the field (5.). Furthermore, by analyzing the impact of these actions, the behavior of the drone swarm may be coordinated to deliver more accurate situational information. Therefore, an automated closed-loop behavior of the drone swarm can be established here. In the end, this intelligent decision assistance for the incident commander would help in extinguishing the fire effectively (6.).

Mission and operational scenario. Figure 4 shows a detailed-level diagrammatic representation of the future FireMan ConOps for managing a drone swarm during the wildfire scenario (i.e., phases 3–6 in Fig. 3). Here, it is estimated that this kind of setup might be operational in approx. Seven years at TRL 8. The Initial Situation depicted in the upper portion of the FSM in Fig. 2 is similar to the scenario and mission description taken into consideration here. It is also assumed here that the diagram is rather self-explanatory as a high-level model of the situation from the human factors' perspective, and no further textual description of the different elements is presented here. The diagram's colors and symbols indicate the following: the orange arrows are interactions and data flows between the different main elements (rounded rectangles) of the entire socio-technical system. The blue color indicates the human operators in the command, control and communications (C3) center and their key outputs (the blue arrows). The main elements of this ConOps include, for instance, the following kinds of properties that were identified relevant for each element. Human operators in a C3 center (see details in Fig. 5): use cases; user needs; tasks and goals; situations and scenarios; and user interfaces. Other actors: incident commander; fire brigades; aerial surveillance flights (traditional aviation); voluntary fire brigades; assistance from authorities; meteorological institute; etc. Operational environment: extent of the fire area; weather conditions; characteristics of the terrain; etc. Semi-autonomous drone swarm: level of automation and sensor fusion; high-quality video feeds; sensors for situational awareness, technical platforms and data communications, and UAS traffic management. All these properties affect the operational activities of the socio-technical system described in our ConOps.

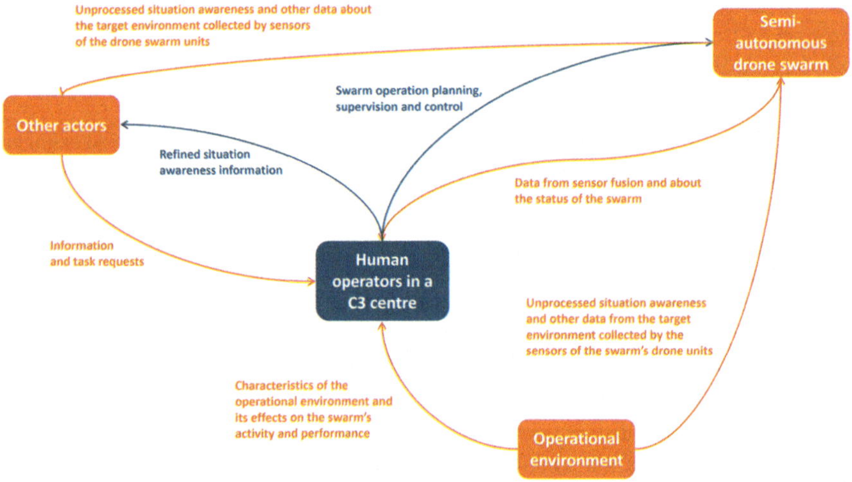

Fig. 4 Final ConOps for a drone swarm during a wildfire situation (i.e., the FireMan ConOps)

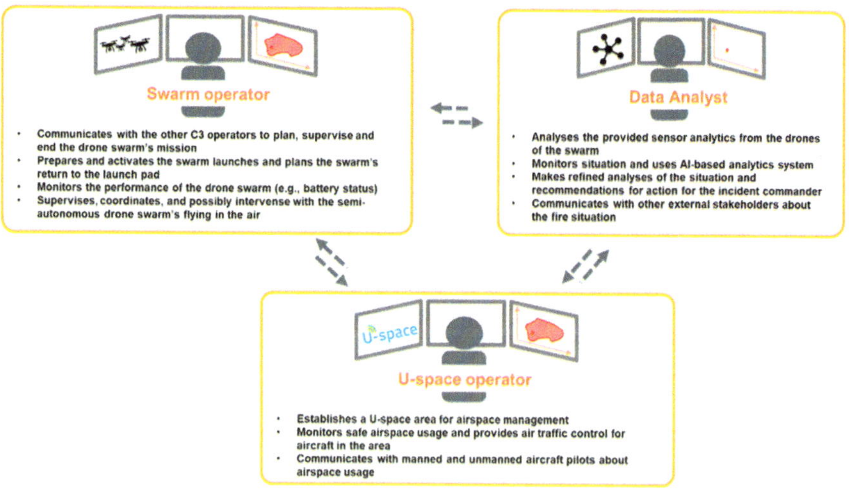

Fig. 5 Human operators and their main roles at a C3 center

System requirements: A semi-autonomous swarm of drones makes up the used technical solution. The individual drones (current reference Parrot ANAFI Ai, see details in Sect. 4.5) can fly for 30 minutes and could be outfitted with a variety of payloads, such as RGB and thermal imaging cameras. The drone swarm unit station, complete with launch pads, is mobile and may be requested to be sent to the wildfire location. In order to plan and coordinate the swarm's mission and operations, the system also has a swarm C3 platform. Additionally, if the autonomous flying of the drones does not proceed as planned, this solution allows human interventions. A communications system built on 5G-Advanced or 6G connectivity offers enough capacity for many stakeholders to receive high-quality sensor data from the system. For the safe management of uncrewed and crewed aircraft in the same airspace, a pop-up UTM system is also used.

Main stakeholders. The organization of the drone swarm-based fire monitoring operation and the associated C3 center is usually the primary duty of public law authorities, such as the fire and rescue departments. However, private organizations could also be able to provide this as a service through outsourcing. Collaboration with the meteorological institute, voluntary fire departments, telecom operators, aerial surveillance flight operators, and other authorities—such as the police—is also crucial.

The human operators at the C3 center include a swarm operator, a data analyst, and a U-space operator. Their main responsibilities and roles are listed in Fig. 5. U-space in this context refers to UAS traffic management (see details on U-space, e.g., in [21]).

System interface. An easy-to-use human-machine interface for the interaction between the swarm operator and the swarm is provided. A step-by-step main task progress example from the swarm operation system's usage perspective is provided next:

1. Using her/his map-based swarm C3 user interface, the swarm operator chooses the region to be monitored by painting the area on the map.

2. The swarm flies to its assigned altitude and disperses to the designated reconnaissance region, optimizing surveillance coverage while preserving communication between the various swarm members and the central control.
3. Some of the swarm members remain to keep an eye on the region in question when they detect a fire, while others keep scanning other places.
4. The area's geospatial data and the real-time data from the drones of the swarm are combined by the decision support system.
5. To direct fire extinguishing efforts in a wildfire scenario, a simulation environment is utilized, which uses a simulated forest environment (a digital twin) that closely resembles the actual environment. This makes it possible to predict the fire's growth across many settings and time periods.
6. In order to make better judgments about extinguishing operations, the fire incident commander can analyze this situation model from various angles and time periods to decide the needed actions.

4.5 Fleet Management

The expert feedback in the evaluation of the first version of the ConOps (see [3]) emphasized the importance of fleet management. Furthermore, questions, such as "how many drones would be needed for the presented scenario", were asked by the experts. Therefore, we decided to conduct a numerical experiment regarding the number of drones required for the swarm operations to be cost-efficient in this kind of a situation.

When multiple UAS are cooperatively involved in performing a task as a swarm, the question of fleet management inevitably arises. Determining how many drones are required to complete the mission, for how long, and how do these and other external parameter values compare to operational constraints, such as battery level and maintenance time (e.g., for battery charging/swapping), is critical to achieving success.

In order to explore this aspect experimentally, we devised a simple scenario (based on the same wildfire scenario as in the FSM's "Initial Situation" in Fig. 2) in which a variable number of quadcopters are tasked with monitoring a point of interest (POI, i.e., the fire front) over a period extending the battery level of individual units, mandating a turnover. A turnover means that the drones must eventually go back to base to recharge and, ideally, should be replaced by others to maintain the POI under observation. As was presented previously, our scenario is inspired by a real-world event in which a wildfire is threatening a settlement: initially (i.e., at the time of detection) the fire front (POI) is 2 km from the UAS base that is co-located with the aforementioned settlement. As a benchmark, we first tested performance for a static event (constant distance between the POI and the base) and then with the wildfire advancing toward the settlement at a speed of 0.25 m/s (0.9 km/h). The drone characteristics were chosen to approximate those of a Parrot ANAFI Ai: $30'$ battery level, a (conservative) cruise speed of 10 m/s (36 km/h) and requiring 2 h to reach full charge (using a linear profile, i.e., a fully discharged battery reaches 50% after 1 h plugged in) [43].

The autonomous drone behavior is as follows: every fixed time interval (tunable parameter), the unit with the highest battery charge is dispatched to the POI if and only if its battery level is sufficient to reach the fire front and return to base with a 10% charge left to spare. At the start of the numerical experiment, all drones in the swarm are fully charged (100%, 30′ battery level). Once airborne, the selected UAS flies directly to the POI near which it will hover to keep it under observation for as long as possible (i.e., until its remaining battery charge forces it to go home, taking into account the 10% reserve). Under these rules, it is also possible for a quadcopter to be dispatched to the fire event, only to be forced to head back to base moments after arriving on site.

Performance was then measured as the fraction of time that the POI was being monitored by one or more UAS, over a two hours' time window. At the end of this time windows, the fire front will be a mere 200 m away from the base in the advancing fire version of the scenario. The parameters the influence of which were under investigation are the time interval between launches and the size of the swarm (i.e., number of drone units). Figure 6 shows an example of how the mission would pan out for a small fleet of 3 drones, launching at 15′ interval. On the left of the figure (A) it can clearly be seen that, following the initial period of constant observation, the swarm is struggling to keep up because, with such a small number of units (and so short replacement cycle), drones do not have enough time to recharge their battery before being sent back to monitor the event. This effect is progressively alleviated by the fire moving closer to the base and proportionally reducing time spent in transit on the right of the figure (B).

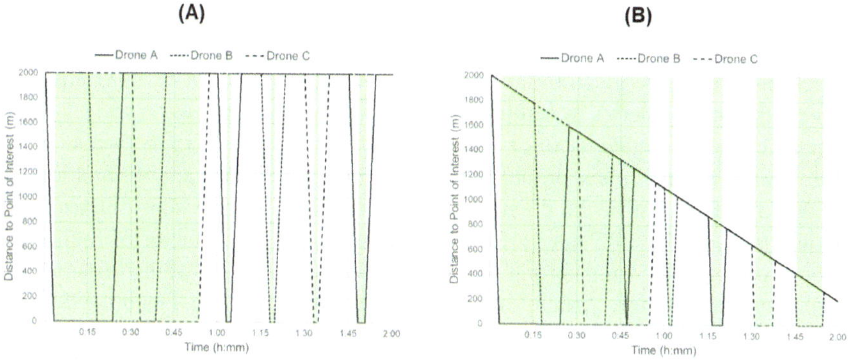

Fig. 6 Visualization of a three units-strong swarm's activity for the static (A) and advancing fire scenarios (B). Both are for a time interval of 15′ between launches. The lines show the movement to and from the POI (fire front) for each drone. The green bands indicate periods during which the POI was being monitored by at least one unit

This performance evaluation method allows us to predict how large the swarm must be in order to keep a certain event under observation by a target number of drones. For instance, the requirement could be to keep a "picket line" of drones hovering 20 m from each other ahead of the advancing fire front, making the necessary number of units concurrently present at any one time a function of the width of the front (100 m → 5 drones, etc.) Figure 7 illustrates how a swarm of 6 (A) and 12 drones (B) would perform, with a launch interval of 15′ and 5′, in the advancing fire scenario (all other parameter values and drone characteristics are identical as in Fig. 6).

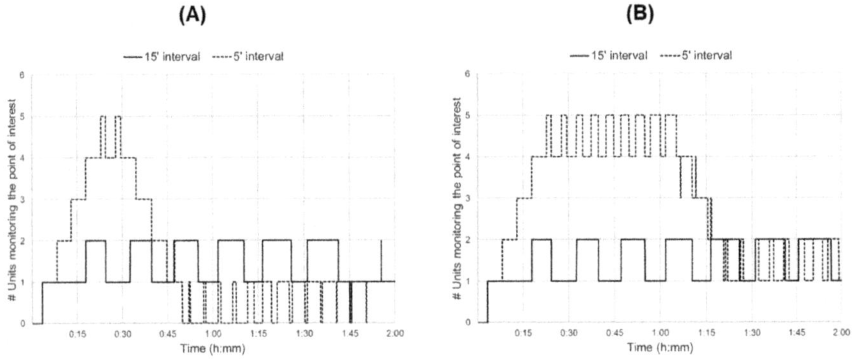

Fig. 7 Evolution of the number of units simultaneously monitoring the point of interest as a function of time, for two different swarm sizes (A: 6 drones, B: 12 drones) and two different time intervals between successive launches ($15'$ and $5'$)

These results clearly demonstrate that it is possible to model and anticipate the collective behavior of a group of autonomous drones tasked with a wildfire monitoring mission, and to control this behavior through two global parameters: number of units committed to the task and time interval between launches. Critically, these two parameters can be set before the mission begins and no more human intervention or specific instructions need to be given to the swarm afterward. However, it remains possible to change mission parameters in real time, should the operator wish to do so.

In the chosen example used for Fig. 7, keeping an interval of $15'$ between launches would allow a swarm of six drones to keep the POI under constant observation by at least one unit for most of the duration of the event, except for a short interruption of coverage (around the 1:45 mark). Shortening the interval by a factor of 3 allows the drones to establish a 5 units-strong picket line in the early stages of the mission, at the cost of considerably more "patchy" monitoring later in the mission. Doubling the size of the swarm and keeping the same time interval between launches ($5'$) would result in the "best of both world", with the drones able to consistently maintain a full picket line for the first hour, then uninterrupted coverage, albeit by a diminishing number of units, after this limit. Further increasing the number of units and/or fine-tuning of the time interval between launches would deliver additional performance gains.

Regarding the question of how many drones would consist of a cost-efficient solution for the fleet, it naturally depends on the problem specifics, such as distance to travel from the base to the event, fire progression speed, and width of the fire front. Ideally, one would want the swarm to be capable of dealing with a variety of cases. However, for the specific example scenario discussed in this paper, 12 drones would be a good approximation. By fine-tuning the interval between departures ($5'$ is a bit short, $15'$ is a bit long), 12 drones would be enough to keep at least a few drones tracking the progression of the fire at all times. If we consider the current price of an ANAFI Ai drone (approx. 5 k€ each), twelve of them would amount to approximately 60 k€ for the whole fleet of drones required for the scenario in question. Naturally, this amount does not include the other system costs mentioned in the system requirements subsection of Chap. 4.4. A major cost would be the drone swarm C3 platform, requiring its own systems engineering process, which's costs and benefits analysis will be presented next.

4.6 Cost-Benefit Analysis

One key feedback from the evaluation of the first ConOps version [3] focused on the usefulness, costs, and benefits of the engineering activities in the life cycle of the suggested swarm platform's engineering process. Therefore, we conducted a cost-benefit analysis regarding relevant human factors engineering activities in making this platform a reality. As stated in Chap. 2, life-cycle cost estimation is connected to the management of risk in the system's life cycle. To quantify the required engineering effort for managing risks, we have developed a spreadsheet-based tool for cost-benefit analysis (see [36]), which is inspired by the work in [44]. The stepwise approach to using this tool was explained in Chap. 3. An example result of the calculation of costs and benefits of systems engineering efforts in the design and development of a semi-autonomous drone swarm platform, as detailed in our ConOps, is presented in Fig. 8. It includes seven phases (the blue boxes in Fig. 8) ranging from identification of risk influence factors to estimation of the total budget to be spent on systems engineering.

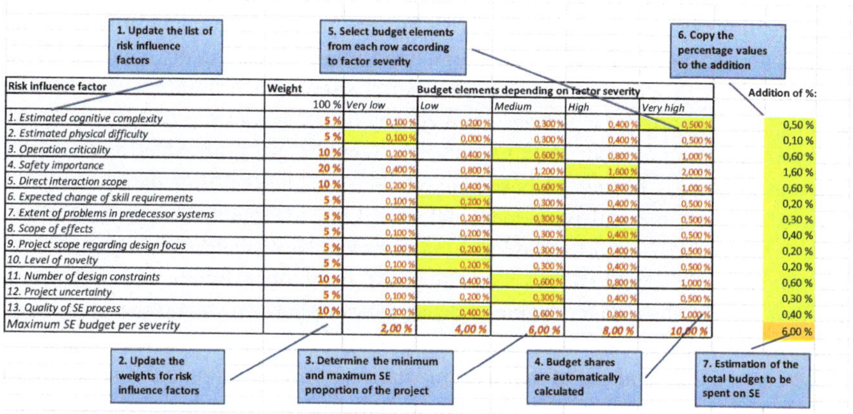

Risk influence factor	Weight	Budget elements depending on factor severity					Addition of %:
	100 %	Very low	Low	Medium	High	Very high	
1. Estimated cognitive complexity	5 %	0,100 %	0,200 %	0,300 %	0,400 %	0,500 %	0,50 %
2. Estimated physical difficulty	5 %	0,100 %	0,000 %	0,300 %	0,400 %	0,500 %	0,10 %
3. Operation criticality	10 %	0,200 %	0,400 %	0,600 %	0,800 %	1,000 %	0,60 %
4. Safety importance	20 %	0,400 %	0,800 %	1,200 %	1,600 %	2,000 %	1,60 %
5. Direct interaction scope	10 %	0,200 %	0,400 %	0,600 %	0,800 %	1,000 %	0,60 %
6. Expected change of skill requirements	5 %	0,100 %	0,200 %	0,300 %	0,400 %	0,500 %	0,20 %
7. Extent of problems in predecessor systems	5 %	0,100 %	0,200 %	0,300 %	0,400 %	0,500 %	0,30 %
8. Scope of effects	5 %	0,100 %	0,200 %	0,300 %	0,400 %	0,500 %	0,40 %
9. Project scope regarding design focus	5 %	0,100 %	0,200 %	0,300 %	0,400 %	0,500 %	0,20 %
10. Level of novelty	5 %	0,100 %	0,200 %	0,300 %	0,400 %	0,500 %	0,20 %
11. Number of design constraints	10 %	0,200 %	0,400 %	0,600 %	0,800 %	1,000 %	0,60 %
12. Project uncertainty	5 %	0,100 %	0,200 %	0,300 %	0,400 %	0,500 %	0,30 %
13. Quality of SE process	10 %	0,200 %	0,400 %	0,600 %	0,800 %	1,000 %	0,40 %
Maximum SE budget per severity		2,00 %	4,00 %	6,00 %	8,00 %	10,00 %	6,00 %

Fig. 8 Calculation of costs and benefits of the Systems Engineering (SE) efforts in the design and development of a semi-autonomous drone swarm system (inspired by [44])

5 Discussion, Conclusions, and Further Research

Our findings suggest that swarm-based drone operations provide an attractive approach to managing wildfires. Drone swarms have the potential to improve wildfire detection and monitoring phases, for instance, because they can provide a more thorough situational awareness by covering a wider geographic region with a variety of onboard sensors and more diverse imaging angles than single-platform drone operations can.

In addition to acquiring the technical components in the implementation of the developed final ConOps, human factors engineering considerations must be made. Swarm operations can be a useful tool for the rescue teams, but they shouldn't take time away from the firefighting crews. Therefore, the system should be user-friendly and have an appropriate level of automation. In an ideal scenario, dedicated human operators could be allocated and the produced information would then help efficiently the extinguishing tasks, which require more direct human involvement. Consequently, drone swarms may also make it possible to manage the resources of the rescue team more effectively. In far-future use cases, drone swarms with the capability of fire extinguishing could also save lives if the number of firefighters on-site at the fire could be reduced in this way.

Based on the findings of our expert studies, a thermal imaging camera provides an efficient way to identify the advancing fire front since it can identify the heat source even through smoke. Extinguishing operations may be planned and carried out appropriately since it provides accurate information on the specific location and progress of the fire front. Fire brigades have utilized thermal imaging cameras to monitor residential building fires and they have shown enormous potential also for wildfire monitoring.

Other surveillance and monitoring mission use cases, which call for the observation of vast terrestrial regions, might also make use of a similar drone swarm solution as the one described in our study. To adapt our solution to other domains, one needs to retrain the AI algorithms to identify other pertinent points of interest (than wildfire) from the ground. The search and rescue of a mission person, border patrolling (including the identification of, for instance, unauthorized people and vehicles), or reconnaissance of enemy positions in a combat situation are a few examples of these additional use cases.

Compared to conventional quadcopters, fixed-wing drones usually have longer operational windows. Therefore, in the future, aerial surveillance flights with crewed aircraft may be supported, or perhaps even replaced, by fixed-wing drone swarms with an effective division of labor. Additionally, this would make it possible for surveillance operations to employ further AI-based analytical tools and real-time data processing.

Based on our numerical experiment, a fleet of 12 quadcopters would be cost-efficient for the scenario envisioned for the FireMan ConOps. Nevertheless, for a future system to be operationally flexible for various situations, even more drones would be required to be a part of the fleet. The developed drone swarm platform should therefore be able to adapt and work with any reasonable number of drones deployed to the swarm.

The cost-benefit analysis of the implementation of the envisioned drone swarm platform is useful to be included in the ConOps documentation, even though a detailed CBA is typically missing from these descriptions. One reason for them to be omitted is that it is very difficult to evaluate the costs and benefits of future systems based on a quite sketchy description of them in the ConOps. However, the CBA can be accomplished at different levels of resolution: by focusing on either to the life-cycle costs and benefits of the proposed new system compared to the previous one or by focusing to the costs and benefits of the engineering work in development phases. Our example presented in Fig. 8 is related to the latter and aims to provide rigidity in this respect.

During this research, we discovered that the utilized ConOps approach is methodologically an effective tool in serving as a boundary object for all the stakeholders involved in the design of intricate socio-technical systems and helps to convey a variety of relevant matters to different stakeholders. A ConOps document provides the means of discussing needs prioritization with experts with diverse backgrounds, particularly when several different types of requirements must be incorporated into the process.

A key scientific contribution of this article has been the demonstration of implementing science-based methods from various disciplines to form a ConOps. To our knowledge, ConOps descriptions are often described from one scientific perspective. Our paper aimed to fill this gap via its various examples on how a combination of descriptions from various scientific disciplines can be potentially done in a ConOps. In detail, we had a combination of human factors (e.g., the FSM), technical (e.g., system requirements), mathematical/numerical experiments (in the fleet management performance evaluation), and economics (e.g., the CBA) aspects considered in the ConOps. To our knowledge this kind of unique combination of multi-disciplinary studies has not been previously conducted in the context of developing a novel ConOps.

In this respect, we see that more systemic cross-disciplinary approaches to increase the usefulness and rigidity of a ConOps are needed. As discussed, some elements of these approaches were included in this paper regarding certain parts of the ConOps. Even more systemic contributions in the future could include, for instance, system dynamics modeling [45] of the non-linear phenomena and their development over time regarding the entire socio-technical system described in the ConOps. In this way, rigorous mathematical models of the target socio-technical system per se could be analyzed and simulated to discover hidden complexities or undesirable tacit phenomena.

On the practical level, the development and application of the high-level ConOps presented in this paper into more feasible real-world solutions will be the main focus of future work. These solutions will involve creating suitable swarming algorithms, simulating the swarming behavior, applying the algorithms into real drone swarm units in a proof of concept, creating machine vision solutions for smoke and fire detection from multiple remote sensing sources, and creating workable operational protocols for the relevant actors. A physical implementation of a smaller-scale drone swarm according to this ConOps on Technology Readiness Level 4 has already been conducted in [46, 47]. In this regard, future work involves the further development of this solution (to TRLs 5–7) and its implementation into an operational solution at TRL 8.

Acknowledgements. This research is funded by the Academy of Finland (project grant numbers 348010, 346710, and 348008) and conducted as part of the unmanned aerial systems-based solutions for real-time management of wildfires (FireMan) project.

References

1. Roldán-Gómez, J., González-Gironda, E., & Barrientos, A. (2021). A survey on robotic technologies for forest firefighting: Applying drone swarms to improve firefighters' efficiency and safety. *Applied Sciences, 11*(1), 363.
2. Yuan, C., Liu, Z., & Zhang, Y. (2017). Fire detection using infrared images for UAV-based forest fire surveillance. In *2017 International Conference on Unmanned Aircraft Systems (ICUAS)* (pp. 567–572). IEEE.
3. Karvonen, H., Honkavaara, E., Röning, J., Kramar, V., & Sassi, J. (2023). Using a semi-autonomous drone swarm to support wildfire management–a concept of operations

development study. In *International Conference on Human-Computer Interaction 2023* (pp. 234–252). Springer Nature Switzerland.
4. Perikleous, D., Koustas, G., Velanas, S., Margariti, K., Velanas, P., & Gonzalez-Aguilera, D. (2024). A novel drone design based on a reconfigurable unmanned aerial vehicle for wildfire management. *Drones, 8*(5), 203.
5. Çoğay, S., & Seçinti, G. (2022). Phoenix: Aerial monitoring for fighting wildfires. *Drones, 7*(1), 19.
6. Akhloufi, M. A., Couturier, A., & Castro, N. A. (2021). Unmanned aerial vehicles for wildland fires: Sensing, perception, cooperation and assistance. *Drones, 5*(1), 15.
7. Bjurling, O., Granlund, R., Alfredson, J., Arvola, M., & Ziemke, T. (2020). Drone swarms in forest firefighting: A local development case study of multi-level human-swarm interaction. In *Nordic Conference on Human-Computer Interaction 2020* (pp. 1–7). ACM.
8. Innocente, M., & Grasso, P. (2019). Self-organising swarms of firefighting drones: Harnessing the power of collective intelligence in decentralised multi-robot systems. *Journal of Computational Science, 34*, 80–101.
9. Pham, H., La, H., Feil-Seifer, D., Deans, M. (2020). A distributed control framework of multiple unmanned aerial vehicles for dynamic wildfire tracking. *IEEE Transactions on Systems, Man, and Cybernetics: Systems, 50*(4), 1537–1548.
10. Ozkan, O. (2021). Optimization of the distance-constrained multi-based multi-UAV routing problem with simulated annealing and local search-based matheuristic to detect forest fires: The case of Turkey. *Applied Soft Computing, 113*(B), 108015. ISSN 1568-4946.
11. Arafat, M., & Moh, S. (2021). Bio-inspired approaches for energy-efficient localization and clustering in UAV networks for monitoring wildfires in remote areas. *IEEE Access, 9*, 18649–18669.
12. Bharany, S., Sharma, S., Frnda, J., Shuaib, M., Khalid, M., Hussain, S., Iqbal, J., & Ullah, S. (2022). Wildfire monitoring based on energy efficient clustering approach for FANETS. *Drones, 6*(8), 193.
13. Seraj, E., Silva, A., & Gombolay, M. (2022). Multi-UAV planning for cooperative wildfire coverage and tracking with quality-of-service guarantees. *Autonomous Agents and Multi-Agent Systems, 36*(2), 39.
14. Munawar, H. S., Gharineiat, Z., Akram, J., & Imran Khan, S. (2022). Framework for burnt area mapping and evacuation problem using aerial imagery analysis. *Fire, 5*(4), 122.
15. Bailon-Ruiz R., Bit-Monnot A., & Lacroix S. (2022). Real-time wildfire monitoring with a fleet of UAVs. *Robotics and Autonomous Systems, 152*, 104071
16. Saffre, F., Hildmann, H., Karvonen, H., & Lind, T. (2022). Monitoring and cordoning wildfires with an autonomous swarm of unmanned aerial vehicles. *Drones, 6*(10), 301.
17. Fairley, R., & Thayer, R. (1997). The concept of operations: The bridge from operational requirements to technical specifications. *Annals of Software Engineering, 3*(1), 417–432.
18. Laarni, J., Väätänen, A., Karvonen, H., Lastusilta, T., & Saffre, F. (2022). Development of a concept of operations for a counter-swarm scenario. In R. Goebel, W. Wahlster, & Z. Zhou (eds.), *International Conference on Human-Computer Interaction 2022*, LNAI (Vol. 10037, pp. 49–63). Springer.
19. Chairman of the Joint Chiefs of Staff: Joint Publication 5–0; Joint Operation Planning. US Department of Defence (2011).
20. Baxley, B., Johnson, W., Scardina, J., & Shay, R. (2016). *Air traffic management technology demonstration-1 concept of operations (ATD-1 ConOps)*, version 3.0 (No. NASA/TM-2016-219213).
21. SESAR 3 JU, CORUS-XUAM: U-space ConOps and Architecture (Edition 4) (2023).
22. Salleh, M., & Low, K. (2017). Concept of operations (ConOps) for traffic management of Unmanned Aircraft Systems (TM-UAS) in urban environment. In *AIAA Information Systems-AIAA Infotech@ Aerospace 2017* (p. 223), AIAA, Grapevine, Texas, US.

23. Cummings, M., da Silva, F., & Scott, S. (2007). *Design methodology for unmanned aerial vehicle (UAV) team coordination.* MIT Humans and Automation Laboratory.
24. Pratt, K., Murphy, R., Stover, S., & Griffin, C. (2009). CONOPS and autonomy recommendations for VTOL small unmanned aerial system based on Hurricane Katrina operations. *Journal of Field Robotics, 26*(8), 636–650.
25. Väätänen, A., Laarni, J., & Höyhtyä, M. (2019). Development of a concept of operations for autonomous systems. In J. Chen (ed.), *AHFE 2019* (pp. 208–216). Springer.
26. Schwandt, T. (2015). *Evaluation foundations revisited: Cultivating a life of the mind for practice.* Stanford University Press.
27. Rodríguez-Sanz, Á., & Rubio Andrada, L. (2023). Cost–benefit analysis of investments in air traffic management infrastructures: A behavioral economics approach. *Aerospace 2023, 10,* 383.
28. Zografos, K. G., & Androutsopoulos, K. N. (2011). Economic and institutional impacts of introduction of self-separation concept of operations in air traffic management. *Transportation Research Record, 2214*(1), 94–102.
29. Mohleji, S. C., Bhadra, D., Ostwald, P. A., & Wieland, F. P. (2005). A process for estimating cost/benefits of future air transportation system operational concepts based on 4D navigation. In *The 23rd Digital Avionics Systems Conference* (IEEE Cat. No.04CH37576), 2004. IEEE.
30. King, J. (2019). *Evaluation and value for money: Development of and approach using explicit evaluative reasoning.* Doctoral dissertation. University of Melbourne.
31. Levin, H. M., & McEwan, P. J. (2001). *Cost-effectiveness analysis.* Sage.
32. Eurocontrol: Human Factors Module. (1999). *A business case for human factors investment.* Eurocontrol.
33. COPE Homepage. http://cope.vtt.fi/project.htm. Last accessed Feb 18, 2023.
34. Joint Authorities for Rulemaking of Unmanned Systems (JARUS): JARUS guidelines on Specific Operations Risk Assessment (SORA), JARUS (2019).
35. Norros, L.: Acting Under Uncertainty—The Core-Task Analysis in Naturalistic Study of Work and Expertise. VTT, Espoo, Finland (2004).
36. Norros, L., Liinasuo, M., & Hutton, R. (2011). Evaluating the potential of new technological tools for safety critical work. *Interacting with Computers, 23*(4), 299–307.
37. Savioja, P., Norros, L., & Salo, L. (2012). Functional situation models in analyses of operating practices in complex work. In P. Turner, S. Turner (eds.), *European Conference on Cognitive Ergonomics 2012* (pp. 58–64). ACM.
38. Vicente, K. (1999). *Cognitive work analysis: Toward safe, productive, and healthy computer-based work.* CRC Press.
39. Laarni, J. (2024). *SAFER2028-TONUS, Measuring HFE program impact*, VTT Research Report VTT-R-00057-24, VTT.
40. Savioja, P., & Norros, L. (2013). Systems usability framework for evaluating tools in safety–critical work. *Cognition, Technology & Work 15*, 255–275.
41. Kramar, V., Röning, J., Erkkilä, J., Hinkula, H., Kolli, T., & Rauhala, A. (2021). Unmanned aircraft systems and the nordic challenges. In T. Lipping (ed.), *New Developments and Environmental Applications of Drones - Proceedings of FinDrones 2020* (pp. 1–30). Springer.
42. Kramar, V., Hinkula, H., Kolli, T., Rauhala, A., Erkkilä, J., & Röning, J. (2021). Overview of the nordic challenges for unmanned aircraft systems. In: *Proceedings of the 30th Conference of FRUCT Association* (pp. 86–98). IEEE.
43. *Parrot: How do I charge the smart battery of ANAFI Ai*? https://www.parrot.com/en/support/anafi-ai/how-do-i-charge-the-smart-battery-of-anafi-ai. Accessed Nov 27, 2024.
44. Bruseberg, A. (2009). *Cost-benefit analysis for human factors integration: A practical guide.* BAE Systems.
45. Forrester, J. W. (1961). *Industrial dynamics.* MIT Press.

46. Coffeng, S. (2024). *Decentralised drone swarm for wildfire detection and monitoring.* Master's Thesis, Aalto University.
47. Coffeng, S., Hildmann, H., Karvonen, H., & Saffre, F. (2024). *Self-organising drone swarms for detection and monitoring of wildfires, submitted for review at the Drones for Resilient Nature—Proceedings of FinDrones 2024*, to be published by Springer in the Springer Remote Sensing/Photogrammetry book series.

Regulatory Frameworks and Policy Analysis

Fragmented Implementation of Regulations for Unmanned Aircraft in Europe?

Anssi Rauhala(✉) and Pekka Leviäkangas

Civil Engineering, University of Oulu, Oulu, Finland
anssi.rauhala@oulu.fi

Abstract. The regulations on unmanned aircraft systems (UASs) changed considerably when Regulation (EU) 2018/1139 on common rules in civil aviation was introduced in the European Union (EU). The new regulation applies to nearly all UAS, regardless of the weight and type of the UAS, and devolved most regulatory powers from member states to the EU Commission and European Aviation Safety Agency (EASA). While the regulations enacted by the Commission (2019/947 and 2019/945) and the guidelines provided by EASA detail the provisions for operating and designing UASs, the implementation and procedural definitions were left to the member states and their nominated competent authorities. This paper investigates the differences in the implementation of UAS regulations in selected member states. Although, to a certain extent, most countries have adopted very similar approaches, there is also variation between the member states in their attempts to address different aspects of the new regulations. Depending on the member state, for example, easily and openly available services and the costs associated with mandatory procedures vary widely. Although nominally operating under the Single European Sky, easy access to airspace may also depend on which country the operation takes place in.

Keywords: UAS · EU · regulation

1 Introduction

The civil applications of unmanned aircraft systems (UASs) have witnessed considerable growth over the past two decades. However, UASs may also pose risks to safety, security, and privacy, for example. To mitigate potential risks and concerns about privacy, the first national regulations started to emerge in 2002–2003 [1]. In the European Union (EU), Regulation (EC) 216/2008 covered only UASs with a maximum take-off mass (MTOM) exceeding 150 kg. This left much of the regulatory powers concerning lighter UASs in the hands of individual member states. Thus, the member states set forth their own policies on the use of UASs [2]. These member state-specific approaches led to fragmentation and resulted in a heterogeneous set of national-level rules. Additionally, the level and quality of implementation varied [1].

The fragmented, dualistic implementation of Reg. (EC) 2008/216 and national regulations was replaced in 2018 by the centralized, top-down framework of Regulation

M. Laakkonen (ed.), *DroneAI for Resilient Nature*,
https://doi.org/10.1007/978-3-032-19008-6_9

(EU) 2018/1139, which applied a risk-based approach to ensure security and safety for aircraft operation in the Single European Sky [3, 4]. The new 'Basic Regulation' gave the EU Commission a further mandate to delegate and implement acts in regulating the UAS sector. The European Aviation Safety Agency (EASA)—comprising all 27 EU countries plus Iceland, Liechtenstein, Switzerland, and Norway as members—is mandated to draft and issue technical regulations upon the request of the Commission. EASA can also introduce standards on how to integrate UASs into the European single airspace strategy. It can coordinate the actions of member states concerning certifications, for instance. It also has soft powers to introduce guidance materials and define acceptable means of compliance. This mandate can be described as the middle ground between top-down regulatory powers and bending to the needs of the market [3].

While market projections promise growth for UAS-related business and services, the market growth is also quite dependent on agile governance that can harness and foster innovation. Although regulators typically resort to the traditional top-down approach, a more flexible bottom-up approach with a flexible operating and business environment is called for by developers, operators, and manufacturers to enable rapid development [5, 6]. Simultaneously, the ever-evolving and changing regulatory landscape erodes the predictability of governance frameworks that are typically seen as a prerequisite to invest in the sector, and it certainly makes market entry efforts more demanding. Even existing operators may have trouble keeping up with the regulations. The State of the Drone Industry Report 2022 pointed out that 42% of respondents in the 2022 report found it difficult to keep track of the changing laws and regulations, compared to 25% of respondents two years earlier, in 2020 [7]. Similarly, in the Global State of Drones barometer, regulatory obstacles have ranked as the biggest challenge for the industry for at least two years in a row (2023–2024) and rule-making authorities have been seen as the most important market-driving factor every year since 2019 [8].

The adoption of the new UAS regulations has not been without issues, including multiple delays at EU and national levels [9]. It is yet to be seen whether the adopted regulatory framework can overall provide either regulatory flexibility or regulatory stability. Furthermore, some problems may yet arise as real-world experiences and data are required to validate the risk-based approach applied to UAS regulations. EASA has acknowledged that there is a lack of data for thorough safety risk assessments regarding the U-space concept. There is also a need to raise awareness about the importance of reporting occurrences [10]. Some authors have raised concerns about the proper balance and issues that remain open regarding related fields such as data protection and privacy, tortious liability, criminal justice, telecommunication, cybersecurity law, and General Data Protection Regulation [3, 4]. For example, there are no harmonized rules on tort liability in general and no specific liability rules for UAS in the EU [11]. It remains to be seen whether insurance companies and policies will dictate additional rules for UAS operations in the future [1]. Although it is too early to evaluate the broader implications of the new regulations, it is possible to assess how the implementation has been handled at the member state level.

As part of the EU Horizon funded SPADE project, we have investigated national differences while trying to pinpoint best practices and some existing challenges in UAS regulation and governance. We have analyzed country-level differences in the adoption of the new rules, including practical aspects from the user/operator perspective, such as whether the English language is available for registration and training. We have also investigated issues of mandatory liability insurance, open access to training and other materials without identification or registration, and some local exceptions, such as permits to use the airspace and registration of operators.

2 Overview of Current UAS Regulations in the European Union

Unlike the previous Regulation (EC) 2008/216, the new Regulation (EU) 2018/1139 on common rules in civil aviation applies to all civil uses of UAS, regardless of their size and weight, with only a few exceptions, such as tethered aircraft with a maximum take-off mass of no more than 1 kg [9]. This new regulation also transfers the main regulatory powers to the European Commission and EASA. However, some exceptions to the centralized model remain. For instance, member states can establish national rules for public security reasons, such as defining prohibited or restricted zones for UAS operations. Additionally, member states can grant temporary exemptions from non-critical requirements due to urgent circumstances or amend implementing or delegated acts when it offers advantages and greater efficiency [4].

The Implementing Regulation (EU) 2019/947 sets out detailed provisions for the operation of UASs, covering both personal and organizational use. In contrast, the Delegated Regulation (EU) 2019/945 is more technical, specifying the design and types of UAS intended for operation as defined by Regulation (EU) 2019/947. Additionally, the Delegated Regulation (EU) 2019/945 imposes rules on third-country operators conducting operations within the Single European Sky airspace under Regulation (EU) 2019/947. These regulations categorize civil UAS operations into "open," "specific," and "certified" categories, establishing boundaries for the operational contexts of each category [9, 12].

The open category outlines low-risk UAS use cases and does not require special permits or operational authorizations [9]. However, operator registration, online pilot training, and examinations are typically required. General requirements for the open category include a maximum flight height of 120 meters, a maximum take-off mass (MTOM) of less than 25 kg, and operations conducted within visual line of sight, maintaining a safe distance from uninvolved people. Starting January 1, 2024, UASs in the open category must also have a class identification marking (C0–C4), be privately built, or have been purchased before December 31, 2023. The open category is further divided into subcategories (A1–A3), each with specific limitations based on the MTOM (Table 1). If the open category requirements cannot be met, a risk assessment and possibly an operational authorization from the national competent authority are needed. Competent authorities can also designate specific UAS geographical zones where operations are exempt from certain open category requirements, such as the 25 kg MTOM limit or the rule that pilots operate only one UAS at a time. The certified category involves high-risk UAS operations, comparable to manned aviation, and requires certified and licensed operators.

With the new basic regulation 2018/1139 applying to all civil UAS regardless of weight and size, amendments were made to Regulation (EU) 996/2010 on the investigation and prevention of accidents and incidents in civil aviation, and Regulation (EU)

Table 1. Subcategories A1–A3 for the open category UAS operations. Not included in the table are legacy UAS purchased before December 31, 2023, and privately manufactured UAS, which can be operated in subcategory A1 when their mass is below 250 g, and in subcategory A3 when their mass is above 250 g but below 25 kg

A1	A2	A3
UAS mass < 900 g	UAS mass < 4 kg	UAS mass < 25 kg
C classes: C0, C1	C classes: C2	C classes: C2, C3, C4
Flights are permitted in densely populated areas above individual people, but not over assemblies of people	Flights are permitted in densely populated areas at safe distances from uninvolved people	Flights are permitted in sparsely populated areas and far away from people and settlements
UAS mass < 250 g: No training/examination requirement UAS mass < 900 g: Online theoretical examination.	Online theoretical examination + additional theoretical knowledge examination	Online theoretical examination

376/2014 on the reporting, analysis, and follow-up of occurrences in civil aviation. For UAS-related incidents, the amendments to Regulation 996/2010 limit mandatory investigations to UASs certified by EASA or to events involving fatalities, serious injuries, or manned aircraft [10]. Similarly, amendments to Regulation 376/2014 restrict the mandatory reporting of occurrences (including less serious incidents that may affect aviation safety) to UAS-related occurrences resulting in fatal or serious injuries, or those involving manned aviation [13]. Although accident reporting is mandatory, many accidents are likely to go unreported [10].

The U-space regulatory package (Regulations 2021/664, 2021/665, 2021/666) from 2021 establishes additional rules impacting the design, manufacturing, and use of certain UAS types [14]. Notably, the 2019 and 2021 Acts by the Commission devolve several powers back to the member states, risking legal fragmentation not only in civil aviation rules but also in their complex interactions with other regulatory sectors [14].

3 Methods

In this study, the focus was placed mostly on open category operations and implementation of related regulations as the information and processes are typically more easily available. Methodologically, our analysis is comparative [15]. We aim to present similarities and differences between countries in their implementation of UAS regulations. The national differences in adoption of the regulations and necessary procedures were compared by analyzing UAS-related sections of the competent authorities' websites, to which hyperlinks are provided on the EASA website under the "Civil drones" section [16] and on the European Drone Company website [17]. In some instances, where the hyperlinks were missing or incorrect, the appropriate websites were located via an internet search engine.

We began by investigating all EASA member states, looking at whether the competent authorities' websites have information available in English, whether the training materials are openly available without registration, and whether the website provides information and links to proper channels regarding the mandatory reporting of UAS-related incidents. The availability of UAS-related information and procedures in English was analyzed by comparing the native language website of each competent authority against their English language version of the website (if any). We also investigated what kind of policy the member states have regarding third-party liability insurance. These aspects are important, for example, from the perspective of potential third-country operators, the openness of information, comprehensive reporting of incidents and occurrences, the subsequent proactive identification of safety concerns, and how the member state addresses insurance and liability issues.

Although websites are not the only channel to convey and inform about legislation, perhaps not even the most authoritative one, they are by far the most commonly used, as they are the only de facto accessible option for operators, be they commercial or recreational. In this sense, our analysis also concerns the accessibility of relevant regulatory information and guidance. Analyzing only the implementation of open category rules and the information provided on websites certainly has some limitations. For example, this kind of comparative analysis does not capture the more substantive side of implementation or the legal backgrounds that might be causing the differences. Hence, our analysis is not causal but descriptive, serving as a starting point for causal considerations. Limiting the depth of the analysis, however, allows us to include all member states. A more in-depth analysis of a smaller subset of member states has been recently published in a book edited by Hartmann et al. [18].

After the first set of analyses, we chose three countries from each of Central and Eastern Europe (Croatia, Hungary, Poland), Northern Europe (Finland, Norway, Sweden), Southern Europe (Greece, Italy, Spain), and Western Europe (Germany, Luxembourg, Netherlands). Some of the countries selected for a deeper analysis were project-related, i.e., the home countries of project partners where more work will be done during the project. The list was then supplemented selectively to have an equal number of countries from each sub-region of Europe using the EuroVoc classification while also ensuring that the largest country of the region was included and sufficient information was available on the websites in a relatively accessible manner.

For this smaller set of countries, we looked a bit more in-depth at what services or information are available in English and which services have costs associated with them. Lastly, we investigated a miscellaneous set of questions related to the services offered and ease of access to them, such as whether electronic identification is possible, whether there are openly available aviation map services, whether there are application forms for specific category operation authorization, and whether the website mentions the possibility of applying for a permissive UAS geographical zone.

4 Results and Discussion

As Fig. 1 highlights, only seven of the EASA member states seem to have all UAS-related information available in English, two of which (Ireland and Malta) have English as an official language. Two member states seem to have no UAS-related information in

English at all. The other 22 member states have at least some information available in English, although the scope varies widely. In this category, some have only the airspace map in English, while others have nearly all the available content also in English.

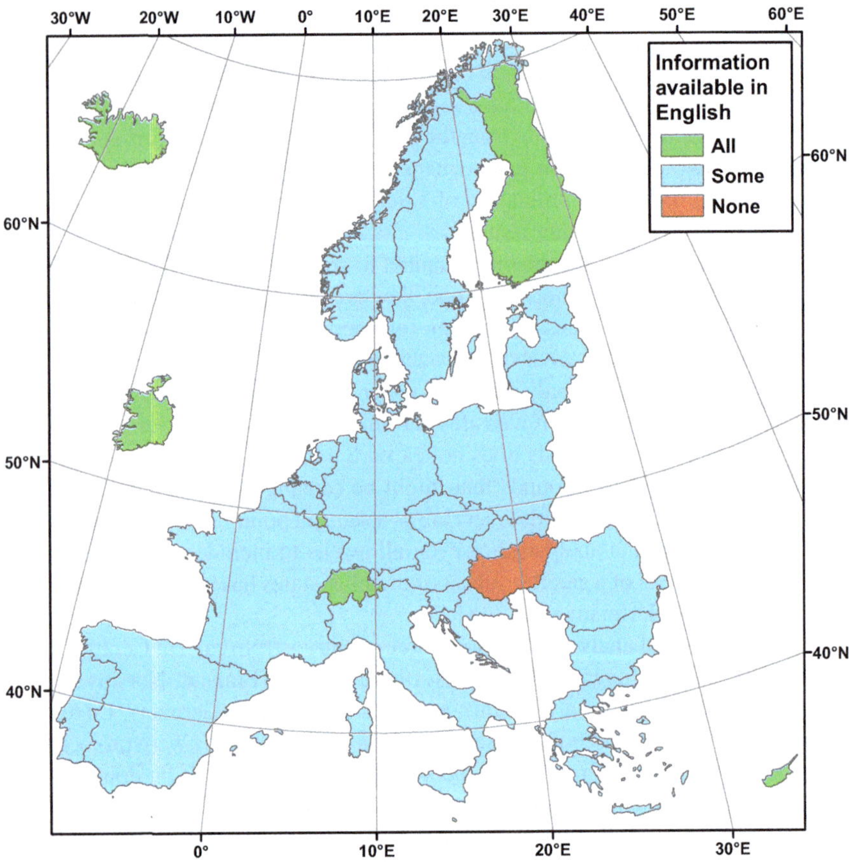

Fig. 1 Map highlighting whether the competent authority website has UAS-related information and procedures available in English as of October 1, 2024. Of the two European microstates that are members of EASA, Malta is in the "All" category and Liechtenstein is in the "None" category

When looking more in-depth into the smaller set of countries, it seems that a map service of airspace zones is generally always available in English (Table 2). Two-thirds of these states have the registration procedure available in English, whereas less than half have training materials or specific category information in English (Table 2). The availability of an English language website is an obvious benefit from the research perspective, but also from the user perspective since according to Reg. (EU) 2019/945, third-country operators need to register in the first EASA member state in which they conduct UAS operations. Although pilot competencies are not tied to the state of residence or to the first state in which a third-country operator conducts UAS operations, the competent authorities acknowledge the need for pilot training in English although they might not offer it. This is seen anecdotally, for example, in the case of Croatia, where the authorities direct non-Croatian speakers to the website of the Luxembourg Directorate of Civil Aviation, where a freely available online training and examination for open category A1/A3 can be accessed by anyone. Regarding the category A2 theoretical examination for non-Croatian speakers, the website notes that "A2 exam from any EU member state will be valid in Croatia".

Table 2 Availability of English language in operator registration, pilot training materials, UAS aviation map service, and specific category instructions. The countries have been grouped according to whether they are in Central and Eastern Europe (CEE), Northern Europe (NE), Southern Europe (SE), or Western Europe (WE)

Country	Registration	Training	Map service	Specific
CEE–Croatia	YES	NO	YES	NO
CEE–Hungary	NO	NO	YES	YES
CEE–Poland	YES	YES	YES	NO
NE–Finland	YES	YES	YES	YES
NE–Norway	YES	YES	YES	YES
NE–Sweden	NO	NO	YES	NO
SE–Greece	YES	NO	YES	NO
SE–Italy	YES	NO	YES	NO
SE–Spain	YES	NO	YES	NO
WE–Germany	NO	NO	YES	NO
WE–Luxembourg	YES	YES	YES	YES
WE–Netherlands	NO	YES	YES	YES

Another important aspect is the availability of information and training materials openly without the need to first register, especially if it requires identity verification or incurs a cost. Openness is often mentioned as a key point in good governance, but one could argue that it benefits everyone if people who have no intention of registering as an operator can also familiarize themselves with the rules. Furthermore, the EASA website warns against fraudulent websites that offer pilot licenses for a price without the need to take an exam. If the website of the competent authority offers little information without the user being required to register first, it is not unexpected for some users to fall victim to fraudulent sites. Considering this, it is a bit surprising that only roughly one-third of the member states offer either all or at least A1/A3 category training materials openly available; the rest of the member states require some kind of registration before access to the materials, or the training has been outsourced to private sector businesses (Fig. 2). Often, the openly available training materials are only in the official languages of the member state.

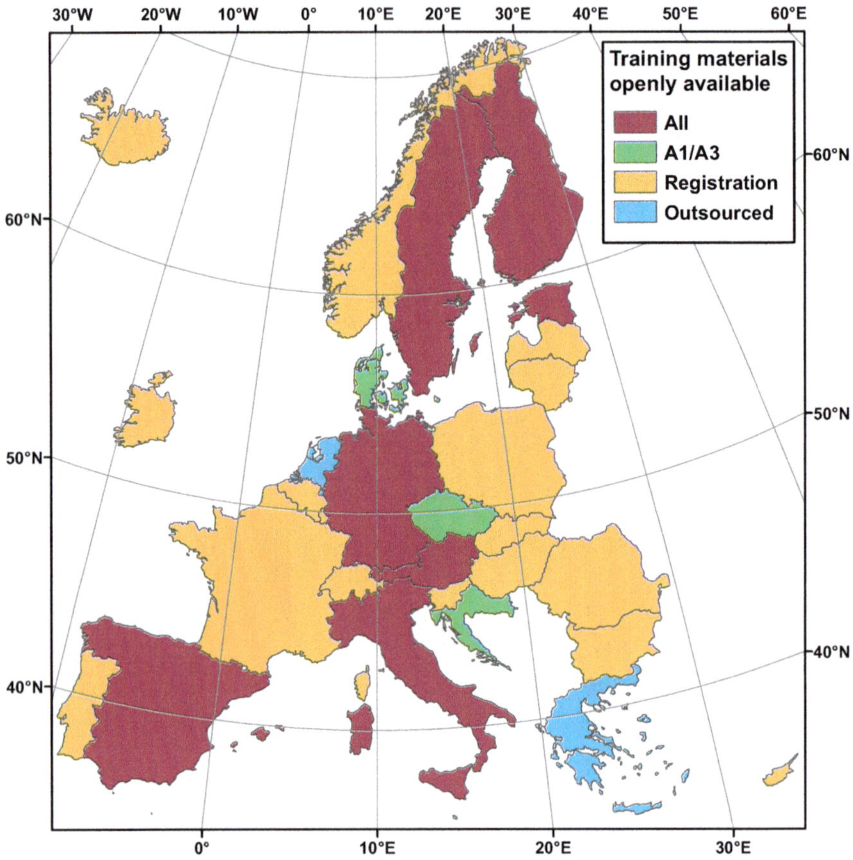

Fig. 2 Map of EASA member states highlighting whether the open category materials are available openly without the need for registration as of October 1, 2024. Of the two European microstates that are members of EASA, Malta is in the "A1/A3" category and Liechtenstein is in the "Registration" category

As mentioned, proactive identification of safety risks requires sufficient reporting of UAS-related incidents and occurrences, and furthermore, reporting certain incidents/occurrences is mandatory. Roughly two-thirds of the member states give either a link for reporting the mentioned incidents/occurrences or both a link and information on reporting incidents/occurrences (Fig. 3). The rest do not seem to have either link or information on reporting available in the UAS section of the competent authority's website. A slight majority of the sites providing both a link and information describe clearly what UAS-related incidents or occurrences must be reported, while the rest only provide general regulations or even less.

Regarding third-party liability insurance, the regulations mandate the operator to have insurance when operating a UAS with a MTOM larger than 20 kg. However, multiple member states have opted to enact stricter rules, with some requiring insurance for all UAS, some for UAS with a MTOM larger than 250 g, and so forth (Fig. 4).

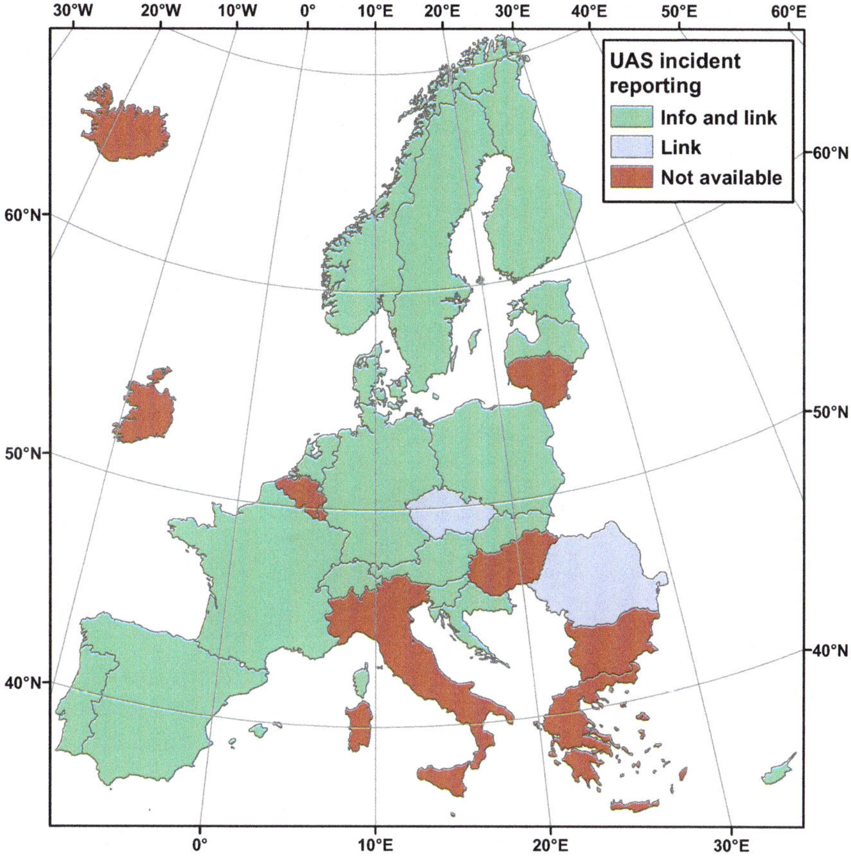

Fig. 3 Map of EASA member states highlighting whether the competent authority website has information on UAS incident and occurrence reporting and/or a link to the reporting service as of October 1, 2024. Note that of the two European microstates that are members of EASA, Malta is in the "Not available" category and Liechtenstein is in the "Info and Link" category

Slightly more than one-third of the member states have adopted the MTOM >20 kg set out in the regulations as their national standard, whereas exactly one-third require liability insurance for all UAS regardless of take-off mass. The rest of the member states land somewhere in between, setting the limit to >250 g, > 900 g, or > 4 kg according to the limits in the open category subcategories and the C classifications for UAS (Table 1).

Looking into the third-party liability insurance requirements (Fig. 4) more in-depth, a bit more nuance can be found. For example, in Croatia, insurance is generally required when the MTOM is greater than 20 kg; however, all aerial work with UAS requires insurance. In Croatia, photography and filming for private purposes are also considered aerial work. In Greece, insurance is always required when the UAS is used for professional use, whereas for private use it is only required if the MTOM is greater than 4 kg. In Spain, on the other hand, liability insurance is mandatory for all UAS regardless of MTOM, but operations in the A1 or A3 subcategories are exempt from the requirement.

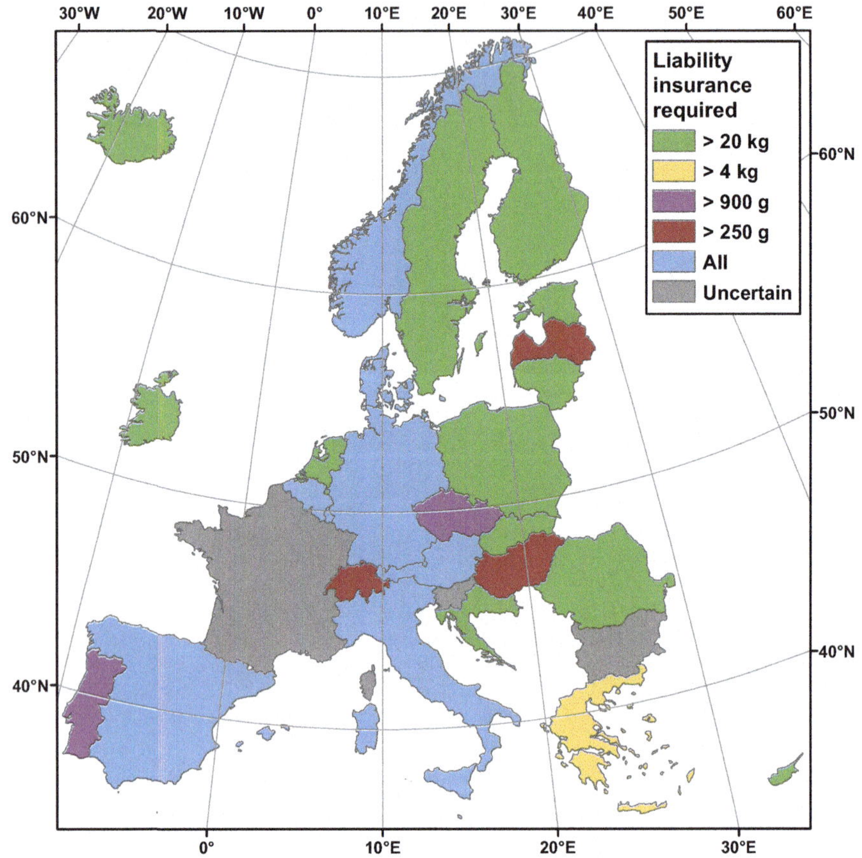

Fig. 4 Map of EASA member states highlighting the requirement for third-party liability insurance based on the take-off mass of the UAS as of October 1, 2024. Of the two European microstates that are members of EASA, Malta is in the "All" category and Liechtenstein is in the "> 250 g" category

Table 3 highlights whether there are costs associated with operator registration, pilot training, or examinations. For Greece and the Netherlands, it may be case-dependent whether a separate fee must be paid for the exams in addition to the training materials. This is due to these states opting to outsource pilot training to private sector businesses. Thus, in some cases, the training and exams may be sold as a package, and in some cases, you may pay separately for A1/A3 and A2 training materials and/or exams. For some member states, such as Croatia, it is not totally clear whether there is a price associated with a certain procedure as the website does not state a price but neither does it clearly state the procedure is free. Some general patterns can be noticed; in the majority of the states, there is a fee for operator registration, whereas training materials are accessible free of charge. When it comes to the examinations, in half of the states, the A1/A3 exam, which can be done unsupervised via the internet, is free.

Table 3 Are there costs associated with operator registration, pilot training, and the A1/A3 or A2 subcategories exams? The countries have been grouped according to whether they are in Central and Eastern Europe (CEE), Northern Europe (NE), Southern Europe (SE), or Western Europe (WE)

Country	Registration	Training	A1/A3 Exam	A2 Exam
CEE–Croatia	No?	NO	No?	No?
CEE–Hungary	YES	NO	YES	YES
CEE–Poland	NO	NO	NO	YES
NE–Finland	YES	NO	NO	YES
NE–Norway	YES	NO	NO	YES
NE–Sweden	YES	NO	YES	YES
SE–Greece	YES	YES	(yes)	(yes)
SE–Italy	YES	NO	YES	YES
SE–Spain	NO	NO	NO	No?
WE–Germany	YES	NO	YES	YES
WE–Luxembourg	NO	NO	NO	NO
WE–Netherlands	YES	YES	(yes)	(yes)

All but one of the twelve member states have some kind of electronic identification available for operator registration (Table 4). In Germany, one must send the authorities a scan of a passport or comparable identification document that will undergo a manual check, which is almost certain to be much more time-consuming than electronic identification. Similarly, all but one of the states have an open map service to check where UAS operations are allowed (Table 4). In Italy, the service requires registration before use. All twelve member states have the forms available to apply for specific category authorization (Table 4). However, the amount of information provided varies significantly. Some, like Greece, provide the bare minimum, while, for example, Germany provides ample information, an 89-page template for the creation of an operations manual, and guidelines and calculation tools for dimensioning flight geography, contingency volume, and ground risk buffer.

The degree to which member states have taken advantage of the ability to set different geographical zones and, for example, limit certain operations also varies. In Hungary, operations over inhabited areas require temporary designated airspace, and registration of the utilized UAS is also mandatory regardless of the EU operator registration. In Poland, each UAS flight should be conducted only after informing the Polish Air Navigation Services Agency about the intention of the flight via the provided web/mobile app. In Norway, non-citizens must notify the authorities if they are flying with a UAS containing any sensors, including a camera. Similarly, in Sweden, dissemination permission is required for aerial photos, videos, laser data, height data, etc. In Germany, which is a federal state, the federated state authorities may, in justified cases, authorize deviations

Table 4 Ease of access and openness of services are highlighted by whether electronic identification is possible, whether there are openly available aviation map services, whether there are application forms for specific category operation authorization, and whether the website mentions the possibility of applying for a permissive UAS geographical zone. The countries have been grouped according to whether they are in Central and Eastern Europe (CEE), Northern Europe (NE), Southern Europe (SE), or Western Europe (WE)

Country	Electronic ID	Open map service	Specific category forms	Permissive UAS zones
CEE–Croatia	YES	YES	YES	YES
CEE–Hungary	YES	YES	YES	NO
CEE–Poland	YES	YES	YES	NO
NE–Finland	YES	YES	YES	YES
NE–Norway	YES	YES	YES	NO
NE–Sweden	YES	YES	YES	YES
SE–Greece	YES	YES	YES	NO
SE–Italy	YES	NO	YES	NO
SE–Spain	YES	YES	YES	(yes)
WE–Germany	NO	YES	YES	NO
WE–Luxembourg	YES	YES	YES	NO
WE–Netherlands	YES	YES	YES	NO

from specified operation conditions while operating in a geographical zone, while the criteria that must be fulfilled depend on the federated state.

The speed at which member states have adopted the new regulations and revised the existing national regulations has also varied widely. For example, in Spain, the use of airspace was, until June 2024, regulated based on Royal Decrees 1036/2017 and 1180/2018, which set out separate requirements for recreational and professional UAS use. Thus, there was, in essence, still a dual approach of old national legislation and the new EU regulations, and only after the enactment of Royal Decree 517/2024 did the general flight rules and UAS geographic zones apply to all UAS operators equally.

5 Conclusions

The regulatory landscape of UAS operations in the European Union has changed dramatically in recent years. The new UAS regulations aimed to harmonize the rules across all EASA member states, but based on our descriptive analysis, the information presented online revealed that there are variations in the implementation of open category related procedures, at least when it comes to the practical aspects from the perspective of UAS operators. The more substantive side of implementation, and a causal analysis of the differences, would require more in-depth research. Similar conclusions and findings are presented elsewhere, too [19]. The findings are not surprising by any means when considering different countries' institutional maturity, governance structures and policymaking cultures.

To a certain extent, most countries have adopted very similar approaches, but the maturity and transparency of implemented processes and rules vary widely. In addition to implementation delays at both the European and national levels, some member states have been slow to adapt to the new rules or to amend existing national legislation. This has, to a degree, led to a sluggish implementation of necessary procedures, or to the overlapping of old national and new EU -level regulations. Many states have opted for more restrictive and prohibitive approaches, while fewer have seized the opportunity to establish permissive UAS geographical zones.

Depending on the member state, the language options, easily and openly available services and information, and especially costs associated with mandatory procedures vary widely. The ease of access to the airspace also depends on which member state the operation takes place in. On some occasions, such as in Germany, different criteria might apply depending on the federated state in addition to the national and EU-level rules.

The variation also presents an opportunity as it provides multiple examples of practical implementation of the new rules and open category procedures. An effort could be made to map out the best practices and implement them throughout the member states. Without easy access to EASA-wide operator registration, pilot training, or incident reporting numbers and data, it is difficult to say whether costs, ease of access to materials and information, etc., discourage potential users or result in incidents or occurrences going unreported. There is also potential here for future studies, including those that aim to identify more precise causal effects.

Some easy steps could be taken to improve the accessibility of information and procedures. Providing an English translation would be an easy solution to make it more user-friendly for third-country operators, or for recreational users (e.g., tourists) and businesses from another member state to gauge any possible national differences. Additionally, making the training materials openly available and free of charge would allow users who have no plans or requirements to register or complete the training (e.g., users operating a small UAS without an installed camera) to easily familiarize themselves with the rules.

Some issues, however, are much more complicated, as they involve related legal and regulatory fields. Liability is often cited as an example of this, as there are no general or UAS-specific rules governing it. This is reflected in the fact that around two-thirds of member states have adopted stricter take-off mass limits for the mandatory third-party liability insurance than those set forth in the general rules. Some member states have even set different rules for the subcategories A1–A3 or depending on whether the operation is commercial or recreational. The lack of harmonization and patchwork of national laws creates uncertainty and makes expanding services EU-wide challenging [11].

Tools to further enable UAS-related business or research and development are available in the regulations, but it is up to the member states and national aviation authorities to implement them. At this point, some member states can be seen as better places for UAS business as their approach is currently more agile, flexible, and permissive. It remains to be seen whether the seemingly reactive governance can ultimately provide either regulatory flexibility or regulatory stability, which are viewed as key enablers of investment and development.

Acknowledgements. This work has been supported by the SPADE project, funded by the European Union's Horizon Europe Research and Innovation program within HORIZON-CL6-2021-GOVERNANCE-01 under Grant Agreement no. 101060778. We thank the editors and two anonymous reviewers for their helpful comments and suggestions that have improved the work.

References

1. Stöcker, C., Bennett, R., Nex, F., Gerke, M., & Zevenbergen, J. (2017). Review of the current state of UAV regulations. *Remote Sensing, 9*, 33–35. https://doi.org/10.3390/rs9050459
2. Tsiamis, N., Efthymiou, L., & Tsagarakis, K. P. (2019). A comparative analysis of the legislation evolution for drone use in OECD countries. *Drones, 3*, 1–15. https://doi.org/10.3390/drones3040075
3. Pagallo, U., & Bassi, E. (2020). The governance of unmanned aircraft systems (UAS): Aviation law, human rights, and the free movement of data in the EU. *Minds Mach (Dordr), 30*, 439–455. https://doi.org/10.1007/s11023-020-09541-8
4. Bassi, E. (2020). From here to 2023: Civil drones operations and the setting of new legal rules for the European single sky. *Journal of Intelligent and Robotic Systems: Theory and Applications, 100*, 493–503. https://doi.org/10.1007/s10846-020-01185-1
5. Du, H., & Heldeweg, M. A. (2019). An experimental approach to regulating non-military unmanned aircraft systems. *International Review of Law, Computers and Technology, 33*, 285–308. https://doi.org/10.1080/13600869.2018.1429721
6. Hernández, G., & Amaral, M. (2022). *Case studies on agile regulatory governance to harness innovation, 0–43*. https://doi.org/10.1787/0fa5e0e6-en
7. DroneDeploy: DroneDeploy's State of the Drone Industry Report 2022. (2022).
8. Drone Industry Insights. (2024). *Global State of Drones, 2024.*
9. Rauhala, A., Tuomela, A., & Leviäkangas, P. (2023). An overview of unmanned aircraft systems (UAS) governance and regulatory frameworks in the European Union (EU). In D. Bochtis, A. C. Tagarakis, & D. Kateris (Eds.), *Unmanned aerial systems in agriculture* (pp. 269–285). Academic Press. https://doi.org/10.1016/B978-0-323-91940-1.00012-8
10. Kasprzyk, P. J., & Konert, A. (2021). Reporting and investigation of unmanned aircraft systems (UAS) accidents and serious incidents. Regulatory perspective. *Journal of Intelligent Robotic Systems, 103*, 3. https://doi.org/10.1007/s10846-021-01447-6
11. Scott, B. I., & Hartmann, J. (2024). International aviation law on liability. In J. Hartmann, B. I. Scott, S. Truxal, A. Bertolini, & A. Masutti (Eds.), *Civil regulation of autonomous unmanned aircraft systems in Europe* (pp. 12–30). Edward Elgar Publishing.
12. Truxal, S., & Scott, B. I. (2024). The regulation of unmanned aircraft systems in the European Union. In J. Hartmann, B. I. Scott, S. Truxal, A. Bertolini, & A. Ma-sutti (Eds.), *Civil regulation of autonomous unmanned aircraft systems in Europe* (pp. 31–63). Edward Elgar Publishing.
13. Konert, A., & Kasprzyk, P. (2021). UAS safety operation—Legal issues on reporting UAS incidents. *Journal of Intelligent Robotic Systems, 103*, 51. https://doi.org/10.1007/s10846-021-01448-5
14. Bassi, E., & Pagallo, U. (2022). A guide to EU drones law, and its work in progress. *SSRN Electronic Journal*. https://doi.org/10.2139/ssrn.4273048

15. Collier, D. (1993). The comparative method. In A. W. Finifer (Ed.), *Political science: The state of the discipline II* (pp. 105–119). American Political Science Association.
16. EASA: Flying in your country—National Aviation Authorities. https://www.easa.europa.eu/en/domains/civil-drones/naa (2024). Accessed Nov 29, 2024.
17. European Drone Company: European drone authorities and no-fly zones. https://europeandronecompany.com/european-drone-authorities-and-nofly-zones/ (2024). Accessed Nov 29, 2024.
18. Hartmann, J., Scott, B. I., Truxal, S., Bertolini, A., & Masutti, A. (Eds.). (2024). *Civil regulation of autonomous unmanned aircraft systems in Europe*. Edward Elgar Publishing.
19. Hartmann, J., Scott, B. I., Truxal, S., Bertolini, A., & Masutti, A. (2024). Conclusion to the legal evolution and regulatory challenges of unmanned aircraft systems in Europe. In J. Hartmann, B. I. Scott, S. Truxal, A. Bertolini, & A. Masutti (Eds.), *Civil regulation of autonomous unmanned aircraft systems in Europe* (pp. 312–316). Edward Elgar Publishing.

The Future of Autonomous Unmanned Aviation: Bridging the Gap Between Technology and Regulation

Jesse Tolvanen(✉), Pedro Henrique Juliano Nardelli, and Tapio Saarelainen

LUT University, Lappeenranta, Finland
{jesse.tolvanen,pedro.nardelli,tapio.saarelainen}@lut.fi

Abstract. This paper presents a Cyber-Physical approach for placing autonomous UAV operations in current regulatory framework with the associated harmonized standards and current international standardization efforts. Literature research revealed the lack of regulatory viewpoints in scientific research in this field. The autonomous UAV operations were divided into 12 categories where two (1: Disaster Management and Emergency Response and 2: Logistics) were selected for further analysis using Cyber-Physical System (CPS) framework. Following the CPS taxonomy, the layers and processes of the modeled operations were mapped to relevant standards and regulations from a technical complexity viewpoint. The technical requirements in conjunction with airspace requirements revealed the complexity and need for both harmonized standardization and implementation of automated Unmanned Traffic Management (UTM). However, the paper recognizes also some issues with full scale U-Space rollout. Another result of the paper is the need for additional Standard Scenarios (STS) which have potential to lower the cost for entry for UAV market. However, the STSs would most likely require new regulatory UAV classification, which in turn requires harmonized standards. The process for creating a new harmonized standard in the EU is a long process, so new Standard Scenarios would be available only in 2030s. For this reason, the paper proposes using an established CPS framework to enable systematic analysis for research on the field.

Keywords: Unmanned aviation · Drone · Cyber-physical system · Standardization

1 Introduction

Unmanned aviation is in a state where the constant development and innovations of the research community and industry are constantly outpacing the regulators and governing bodies [1]. As Unmanned Aerial Vehicle (UAV) technologies continue to develop and new applications are proposed, drones are being designed to solve a multitude of real-world problems. Although it is good for the scientific community to innovate and publish high-flying ideas for drone usage, the sometimes populistic approach might lead to unrealistic expectations about the speed of adoption and implementation of these technologies.

M. Laakkonen (ed.), *DroneAI for Resilient Nature*,
https://doi.org/10.1007/978-3-032-19008-6_10

The literature research conducted for this article reveals that only a small portion of scientific papers on UAVs or drone swarms took the regulatory considerations and limitations into account. This paper aims to describe some of the challenges of introducing autonomous UAVs into airspace from the point of view of regulation and standardization. This paper

1. Explores the use of Cyber-Physical System (CPS) Framework in analyzing the technical complexity, internal, and external dependencies of autonomous Unmanned Aerial Vehicles and Systems and.
2. Using the CPS as a tool to recognize the relevant and critical points reflects on the current state of the UAV regulations and standards.

The research focus of this paper is partitioned into assessing the airspace and technical requirements necessary for autonomous operations using a CPS framework and demonstrating its use as a common reference as a system-level demarcation tool.

Well-publicized collisions and near misses [2–4] in the mid-2010s between drones and airplanes near airports provoked a public debate [5] about the safety risks posed by drones. In 2016, European Union Aviation Safety Agency EASA 'Drone Collision' Task Force assessed the risks of collision between manned and unmanned aircraft [6]. The following years saw the European Commission and EASA work toward harmonizing drone regulations, driven by the need to mitigate the risks caused by the increasing number of drones [7].

2 Autonomous UAV Operations

Although drones act as an enabler for a multitude of new services, the bottleneck for massive scaling is still most often the need for humans to operate UAV systems. An Unmanned Aerial System (UAS) encompasses the UAV and its associated command and control systems.

Depending on the type of mission, even the current Commercial Off The Shelf (COTS) UAVs can fly whole missions by themselves from take-off to landing.

Incorporating additional automation could reduce the need for a human operator if tasks like battery and payload exchange, along with communication with other aviators and air traffic control, are automated.

There are many definitions of the term *autonomy*. The European Union Aviation Safety Agency EASA on their *Easy Access Rules for Unmanned Aircraft Systems* [8] recognizes that there is not yet a common understanding or definition of the of autonomy. EU's drone regulation *EU 2019/947* defines autonomous operation as "*an operation during which an unmanned aircraft operates without the remote pilot being able to intervene*" [9].

Rather than being a Boolean yes/no definition, this paper uses the definition of the autonomy as a spectrum similar to [10]. It can encompass anything from formation control and task allocation of a hundred UAV swarm to activating a Return To Home (RTH) of a single drone on a loss of control link.

2.1 Airspace

Any aircraft that is not on the ground is in some kind of airspace, either controlled or uncontrolled. For manned aviation, it is routine to move between and operate in different airspaces with different requirements and limitations. Our literature research and prior experience from the field show that operating civil UAVs in controlled airspaces faces some challenges.

International Civil Aviation Organization ICAO has classified airspace into seven classes A to G [11]. Classes A to E are controlled airspaces, and classes F and G are uncontrolled.

Operation in controlled airspace requires communication and coordination with Air Traffic Control (ATC), which oversees the use of airspace and, in most cases, provides separation between aircraft [12].

UAV operators need a mechanism to communicate with the air traffic control. While the ATC might be able to provide service to the current level of UAV traffic, the fully scaled and autonomous UAV systems would increase the ATC's workload significantly.

Unmanned Traffic Management (UTM) is a concept of managing UAV traffic fully autonomously or assisted by a human operator [13]. *Commission Implementing Regulation (EU) 2021/664* Article 2 defines the "'U-space airspace' as a UAS geographical zone designated by Member States, where UAS operations are only allowed to take place with the support of U-space services" [14]. The concept of U-Space is an umbrella under which various services and technologies are being developed. U-space can facilitate complex air operations such as Urban Air Mobility which includes transporting cargo and people.

According to Single European Sky ATM Research (SESAR), U-Space is expected to be in an initial implementation phase until 2030 [15] as the experience of UAVs and U-Space grows. Each U-Space may have a unique set of services and have multiple or different service providers. However, there is a minimum set of services that U-Spaces must provide, per [16] defines mandatory U-Space services that subject all operations in an U-space. These services are the network identification service, the geo-awareness service, the UAS flight authorization service, and the traffic information service.

2.2 Technical Complexity Analysis

The total technical complexity of an autonomous UAV system or a swarm is a combination of all subsystems. The UAV-part (the actual flying machine) is mostly mature technology at this time.

The autonomous UAV system can be modeled as a Cyber-Physical System (CPS). Deconstructing a system using a systematic CPS framework enables comparison between different systems in a harmonized and objective manner.

The authors in [17] present a multilayer structure of an intelligent transportation system. Using a CPS framework presented in [18], the system is divided into three layers and the processes under them. The first layer is the physical layer, which includes the sensing and measuring process. Next layer is the data layer with the informing process. The last layer is the decision layer with the acting process.

Physical layer consists of the mechanical components of the UAV such as motors, propellers, battery and frame, but also sensors, radio links, antennas, onboard computers, lights, and other payloads. Physical phenomena such as aerodynamics, radio propagation, and losses are part of the physical layer.

Sensing and measuring process maps the parameters between the physical and data layers. For example, in an UAV context, the sensing might be capturing an image with a thermal camera. In addition to the directly UAV-specific actions, the sensing and measuring includes the system's other sensing tasks, such as radar measurements.

Data layer includes all digital data and information, machine learning and computer vision algorithms, situational awareness, and management tools and systems.

Similarly to the sensing and measuring process, the *Informing process* maps or moves the information between layers. The purpose of the informing process is to feed information from the data layer to the decision layer. Depending on the agent at decision layer, the informing process can take appearance as an user interface for a human, or as an Application Program Interface (API) as an example.

Decision layer includes the processes that act or lead to actions from the information from the data layer and the information processes. The actors in the decision layer can be human operators, or they can be algorithms.

Acting process is the final process, an action taken by the agent following the decision layer.

The more higher level (the physical being the lowest, and decision the highest) things are considered, the more complex the system integration becomes as more stakeholders are needed to coordinate the actions. The layers are always built on top of the lower level layers, which provide some kind of interface or process to transfer the information to and from the higher layers. The Cyber-Physical systems may also be nested. A swarm of drones does contain multiple UAVs, which can themselves be analyzed as a CPS.

Regardless of the operation or mission the autonomous UAV system is performing, the end goal is an action in the physical world. A Cyber-Physical System describes the relation of the real physical world and the cyber-domain. Each layer of the CPS have their own complexities: the lower layers (physical) tend to be more technical, while the higher layers (decision) have more complex socioeconomic challenges and associated questions such as the public acceptance of autonomous UAV swarms and autonomous decision making.

3 Standards and Regulations

The European Union Aviation Safety Agency (EASA) has limited regulatory power within the EU. Its legislative power is exercised through opinions, which can lead to EU regulation or by decisions which can include, for example, the acceptable means of compliance for a specific regulation [19].

The European Commission gave the European Union Aviation Safety Agency EASA a mandate to draft specific regulations on drones by publishing the *Regulation (EU) 2018/1139 on common rules in the field of civil aviation and establishing a European*

Union Aviation Safety Agency. The regulation introduced essential requirements for drones and laid the basis for which the European Commission and EASA drafted the UAV regulations (EU) 2019/945 and (EU) 2019/947 [20].

With the regulations, standardization acts as an enabler of autonomous UAV operations. In general, standardization is—or should be—a market-driven effort to ensure interoperability and a level playing field for commercial entities participating in the same market segment. In aviation, standardization also has additional roles that increase safety and make international operations possible.

The European Union UAV regulations [21, 22] lay the basis for the safety requirements of UAVs, and thus lay the basis on the standardization work conducted by CEN [22].

While (EU) 2019/945 defines the requirements for C-marked UAVS from C0 to C4 which are to be operated in the 'open' category, (EU) 2020/1058 adds the classes C5 and C6 which are to be used in 'specific' category using *Standard Scenarios* (STS). Operations in 'specific' category conducted outside of STS, which means using the Predefined Risk Assessment (PDRA) or Specific Operations Risk Assessment (SORA), does not require the use of a C-marked UAV.

The specific technical requirements of UAVs under SORA and PDRA are left undefined by the regulation (EU) 2019/945. The regulation (EU) 2019/947 sets the risk assessment and performance requirements for the UAVs operated under 'specific' category. Operations performed in this category require operational authorization from a competent authority [9, 21].

Obtaining an operational authorization for any UAV operation under 'specific' category requires the operator to perform a risk assessment. In case of PDRA, the risk assessment is in the form of a table with PASS/FAIL criteria. If the operator can prove that all criteria have been met, the risk assessment is sufficient.

In the case where PDRA is not sufficient, for example, the size of the UAV is larger than what the PDRA allows or the system is autonomous, the risk assessment must be done using the SORA method.

Standards for UAVs are currently being prepared around the world. From a European point of view, the most relevant standardization bodies working on UAV standards are CEN and ISO.

ASD-STAN has published the standards for class C0 to C4 class UAVS. The standards EN 4709-001 to EN 4709-004 cover topics from product requirements and verification, remote identification, and geo-awareness requirements to lighting requirements. CEN technical committee *CEN/TC 471 "Unmanned aircraft systems"* is currently working on the C5 and C6 class UAV standards EN 4709-005 to EN 4709-008.

The International Standardization Organization ISO develops UAV related standards under *ISO/TC 20/SC 16 "Uncrewed aircraft system"*. They have published standards regarding, for example, UAS Traffic Management (ISO 23629), Unmanned Aircraft Systems (ISO 21384), Unmanned aircraft electric energy systems (ISO 24352) and Vertiports (ISO 5015). The ISO is currently developing standards for using hydrogen as an energy source, as there are currently two draft standards *ISO/CD 25009 Unmanned aircraft systems—General requirements and test methods for the hydrogen fuel gas pipes of gaseous hydrogen fuel cell powered UAV* and *ISO/CD 25013 Unmanned aircraft systems—General requirements and test methods for the attachable hydrogen cylinders of gaseous hydrogen fuel cell powered UAV* under development [23].

In the EU, if there is a need for a common standard, the European Commission requests and gives the mandate for the standardization bodies to start the work on a harmonized standard. The mandate has a limited time and scope in which the standard must be completed. For the standard to be harmonized, the reference of the standard must be published in the Official Journal of the European Union (OJEU).

Of the ASD-STAN and CEN UAV standards, currently only EN 4709-002 (Remote Identification) has been published in the OJEU [24].

4 Autonomous UAV Applications

The scientific literature on UAVs often lists various applications in which UAVs and UAV swarms are used or could be used. The authors in [1, 13, 25] have reviewed the literature and categorized UAV applications into various categories.

Based on the publicized literature, we have recognized 12 main categories of UAV applications.

The categories are further divided into two classes, depending on whether the swarming was specifically mentioned or not per literature source. The results are presented in Table 1.

Table 1 Proposed applications of UAVs and a swarm of UAVs found in scientific literature

Application	Use of a UAV	Use as a Swarm
Healthcare and medical	[13, 26, 27]	
Disaster management and emergency response	[1, 13, 28–30]	[1, 13, 27, 31–34]
Public safety and security, surveillance	[1, 13, 35]	[1, 26, 27, 32]
Traffic monitoring and management	[1, 36]	[1, 27, 31]
Delivery and logistics, transportation	[1, 13, 26, 27, 30, 33, 37–39]	[31, 40]
Agriculture	[1, 13, 26, 35, 36]	[31, 33, 41]
Environmental monitoring and remote sensing	[1, 36, 41]	[1, 31]
Industry and infrastructure inspection	[1, 13, 26, 33, 36, 40, 41]	[42]
Land surveying and geospatial mapping	[1, 13, 26, 29, 35, 36, 41, 43]	[27, 32, 40]
Communications (5G etc.)	[1, 13, 41]	[1, 27, 31]
Entertainment and media	[1, 32]	[44],
Military, government	[1, 13, 35]	[26, 27, 32, 33]

Table 1 shows 11 of the 12 categories of UAV applications that have at least one reference from the literature to use UAVs as a swarm for the application. The sources chosen for the table are either sources already in this paper or gathered by using a separate database search. From the application categories, *Disaster management and Emergency Response* (DMER) and *Logistics* were chosen as the basis for the analysis of technical complexity. The chosen UAV applications present two different sets of prerequisites and requirements for the operation concept, relevant regulations, and standards.

The DMER application category includes missions such as search and rescue operations to find missing or lost people [25], enhancing situational awareness in natural or other disasters, monitoring and managing forest fires [13, 34], transporting medical aid to victims in difficult to reach places [13, 27] providing emergency communications via UAV-mounted base stations or radio repeaters [25] or use of drones as a remote sensing platform [1]. There are some overlap with other categories such as medical and logistics, as presented in [30].

UAV-based logistics can be divided into last mile deliveries and for long-distance freight operations. The last mile delivery by drones has already been proven to be feasible by multiple commercial companies [33, 37]. UAVs can be used to deliver daily consumer goods such as food and beverages, healthcare payloads or postal deliveries [26, 38]. This paper focuses only on the last mile delivery UAV logistics.

4.1 Technical Complexity

The autonomous UAV system or swarm, considered as a Cyber-Physical System (CPS), can be analyzed by disassembling it into layers and processes. This section will include some of the existing relevant standards or regulations as examples. Like any complex system, the UAV system or swarm represents a system of systems, where distinct CPS boundaries can be delineated. The purpose of this section is to illustrate the application of the framework and methodology in the analysis of systems.

Physical Layer

Given that DMER operations typically involve time-sensitive tasks, the UAV components of the UAV have the requirement of being robust and simple to facilitate quick field deployments. Similarly, the UAVs used in logistics applications would be expected to run as much of the time as possible, the mechanical durability and interchangeability of the parts may be preferred over optimizing for the lightest weight possible even if using lighter parts might increase the available flight time.

The physical layer also benefits from the standardization of parts, as large scale DMER or logistics operations might require combining equipment from multiple operators. Managing maintenance and spare parts is easier, if the UAV's components have some level of compatibility. The same holds true for batteries, charging infrastructure, and payloads. Having standardization across devices helps with training and improves safety.

The physical layer is already subject to various regulations and EU directives, such as Electromagnetic compatibility (EMC), Low Voltage (LVD), Radio Equipment (RED), Restriction of the use of certain hazardous substances (RoHS) directives [24]. The directives may ensure that the devices are safe to use and not harmful to the environment, but do not guarantee or demand interoperability.

Sensing Process

In disaster management and emergency response operations, the information from the UAVs is produced through some kind of sensor. The raw data from the sensor in an autonomous system need to be transformed, analyzed, or transmitted in order to gain a value from the data.

In any search operation, the performance and capabilities of the sensor payload might limit the way the UAV is operated. For example, the camera's pixel density, frame rate or contrast ratio might limit the autonomous system's ability to recognize a person or an object through a smoke or from a high altitude.

A common way to quantify a camera's performance is to use the Detection, Recognition and Identification ranges (DRI) [45]. An example of a standard in which the test objects for DRI are defined is NATO STANAG 4347 [46].

In addition to the sensor payload, the UAV is sensing its environment. An autonomous system needs information about the weather, the presence and location of other aircraft and about the performance of the UAV's subsystems. In a non-autonomous UAV operation, dangerous weather phenomena such as icing, heavy precipitation, or strong winds are detected and mitigated by the human pilot. Autonomous systems need to rely on sensor data and forecasts.

The conditions at the ground control station can vary quite dramatically to the UAV's position, so the UAV needs to sense the weather by itself unless reliable and up-to-date weather information or forecast is available from an external source. Operations in an U-Space might be able to use the U-Space services, which can include weather data [15].

Data Layer

The UAV system produces large amounts of raw data through the sensing process. In disaster management or emergency response operations, autonomous UAVs supplement the situational awareness of the first responders. The value of the autonomous system for the rescue personnel depend on the accuracy and understandability of the information the system is producing.

As the data layer contains the algorithms and software tools for the autonomous system management, information processing and sharing, the data layer has some regulation in the EU such as the General Data Protection Regulation (GDPR) and the EU's recent Artificial Intelligence (AI) regulation (EU) 2024/1689.

Informing Process

The informing process exists for transforming and transferring information from the data layer to the decision layer. In an autonomous system, it acts as a filter: ideally the process transfers only the information that is needed in the decision layer and the acting process.

In DMER operations, if the informing process presents too much information to the operator, or information he/she does not need, the important information might be missed. Similarly, if the informing process does provide too small amount of data, the situational awareness might deteriorate.

For a fully autonomous drone swarm, the informing process feeds the information to the swarming algorithm. Depending on the topology of the swarming algorithm (centralized versus distributed), the UAVs of the swarm must inform for the system its current location and other information that the algorithm needs. However, the radio spectrum is a limited resource, so the transmission must be efficient. The informing process is a limiting factor for the real-world swarming implementations. A common standard for transmitting information between a UAV and a control station is MAVLink [47]. The protocol is efficient, but it is not specifically designed for UAV swarms and distributed swarming algorithms.

Decision Layer

A fully autonomous UAV system or a drone swarm needs an algorithm to control the movements and the objectives of the UAVs. In the case of distributed control or swarming algorithms, each of the UAVs participating in the swarm executes its own decision process based on its own state, the status of the swarm and the current mission objective.

In disaster management, the mission objective might be locating and analyzing the state of people and property in the mission area, locating forest fires or monitoring an oil spill. There are many swarming algorithms and methods for different mission types, but a common feature in most of the algorithms in the literature require the autonomous agents (UAVs) to have some level of knowledge of the other UAVs.

As discussed in the informing process, the radio communication must be efficient. With a large swarm with tens or hundreds of drones spread over tens of square kilometers, it might be a challenge for all of the UAVs to be informed of each other's locations, as the swarm's size might prevent the farthest of the UAVs to hear each other's transmissions.

A centralized control of UAVs would require the UAVs to be either within the control station's radiolink range, or the UAVs would have to form a mesh network. In case of a natural disaster, the terrestrial cellular network might be unavailable or congested, and thus not be a viable option for the control link.

Acting Process

The acting process is the process of acting based on the decision that was formed through the information process. The act can be the UAV system operator changing the mission, adding a new operation, or it can be the swarming algorithm changing the formation, adding, or removing UAVs from the swarm. It includes also the UAV's autopilot making course and attitude corrections keeping the UAV in the right position and moving in the correct direction.

From the UAV system's or swarm's viewpoint, the acting process takes only the information through the informing process into account if the system is analyzed purely from the CPS viewpoint.

In these operations, there is always a human in the loop, so the acting process which concern the human operator allows other information which is not counted for by the CPS to enter the action and decision process.

A fully autonomous UAV system with distributed control needs to process the decisions and act on the edge, on the UAVs. The level of autonomy a system or UAV is allowed to have depends not only on the level that technology allows, but also on the rules, laws, regulations, and operation authorizations of the UAV operator.

The autonomous operation of any kind of UAV still requires human monitoring. The ISO standard *ISO 21384-3:2023 Unmanned aircraft systems Part 3: Operational procedures* sect. *9.4.5 Autonomous Operations* mandate that all operations are to be monitored so human intervention is possible in case of any safety critical failures that the autonomous system cannot resolve by itself.

4.2 Airspace Requirements

As an emergency or disaster can strike almost anywhere, there is not a single characteristic airspace or operational area that could sufficiently describe or cover all operations in the disaster management and emergency response category. However, in these operations, UAVs do not necessarily need to operate at the same altitude as manned aircraft (> 500 ft), as the performance of the payload of the UAV could limit practical flying altitude [29]. An exception is fixed-wing UAVs that might benefit from operating at higher altitudes.

This means that there is rarely a need to operate outside of uncontrolled (Class G or F) [11] airspace from an altitude perspective.

If a UAV operation is conducted near an active airport, Air Traffic Control (ATC) may need to temporarily close or restrict access to the airspace to allow safe UAV operation. The authors of [48] propose a pop-up U-Spaces or air corridors as a possible solution to the problem. The pop-up U-Space could also be used to coordinate the UAV traffic between multiple operators and UAV service providers, as well as keeping the other aviators informed about the position of the UAVs.

Urban UAV search and rescue operations are challenging as buildings, trees, and other obstacles can prevent the UAV operator from seeing the UAV. Flying Beyond Visual Line Of Sight (BVLOS) requires airspace coordination with other aircraft. Depending on the jurisdiction of the operating country, BVLOS in urban areas might or may not be possible for all drone operators. However, first responders may receive special authorizations to operate BVLOS, or in otherwise restricted areas [1].

For example, in Finland, the police is allowed to waive some of the civil aviation rules such as the VLOS requirements by following the state aviation rules [49].

Using UAVs for last mile logistics brings UAVs close and into the populated areas by design. Urban Air Mobility also known as "drone taxis" will play a role in the future airspace of cities, as autonomous vertical take-off and landing (AVTOL) aircraft will transport people as part of the transportation infrastructure [12].

The movement of people and freight in urban airspace will benefit or even require the implementation of U-Space.

The operational authorization for the logistic UAVs will require the use of 'specific' category as defined in [21]. Autonomous logistics operations using SORA have been studied in [50]. The use of heavy payload UAVs might be limited for use in sparsely populated areas and outside of busy airspaces [51].

5 Discussion

The Cyber-Physical System framework was applied in chapter *Technical complexity* to an autonomous UAV system. The chapter offered a *high-level* overview of the use of the methodology. The limited scope and extent of this paper prevent a fully comprehensive or exhaustive analysis and disassembly to be done for the system. However, the analysis showcased the utility of CPS demarcation. The use of an established framework for system-level analysis provides uniform terminology and methodological demarcation, which could help facilitate comparison and collaboration among various research and regulatory bodies working on various regulatory and standardization work items.

As discussed in chapter *Standards and regulations*, the UAV operations in 'specific'-category can be operated under STS, PDRA or SORA. In the future, where the lower parts of the airspace are going to be used by more operators---electric aviation and UAV automation being the enabling technologies---the operational authorization for autonomous UAV operations will have to become more automatized. To ensure that all operators operate in the airspace in a similar manner, additional Standard Scenarios (STS) would be beneficial. For example, UAV-based logistics would benefit from an STS which would set clear guidelines for all operators, cities, U-Space service providers, and other aviation stakeholders of what to expect from UAV-based logistics. The new STS would not have to limit any company's operation or the use of technological innovations as SORA is always an option, but an STS would reduce the cost to enter the market for new operators.

For the new STS's to have maximum effect, a similar approach to the current STS's that do require C-marked UAVS would be beneficial in ensuring safe implementation of the UAV-based logistics. For the DMER the STS's would likely not have a major effect as the first responders could operate with more specific rules based on per-user exceptions or by using closed airspaces.

For logistics UAVs to have C-marking, requirements need to be developed through the harmonized standardization path. The timeframe from the European Commission giving the mandate to a standardization organization to publishing the standard will take many years. It can be expected that the new STS's with the corresponding standards will be in force around 2030. Before this happens, UAV operators need to apply for the operating license through other means.

If multiple operators are to use the same airspace volume at the same time using heterogeneous UAVs: fixed wing, rotary wing, or even blimps or balloons, a safe and automatic air traffic management must also be implemented.

The use of U-Spaces is a two-bladed sword: the overuse of U-Spaces would hinder the recreational and open-category uses of the UAVs as the operation would require utilizing the U-Space services. As the U-Space service providers would most likely be at least in part commercial companies, the cost of UAV aviation would go up.

However, U-Spaces would make more commercial applications viable as UAV companies could outsource some of the required actions such as remote-ID and weather services to the U-Space service providers (USSP).

The EASA's CONOPS for U-Spaces is a good start for providing a common reference for the new and existing USSPs, UAV manufacturers and the UAV companies of what to expect from the U-Spaces at the moment.

6 Conclusion

The article introduced 12 UAV application categories, from which two applications were analyzed from airspace requirements and technical complexity points of view. The U-Space and UTM in general were recognized as an enabler for future logistics operations. Disaster management and emergency response operations could benefit from more flexible U-Spaces in case a lot of unmanned traffic would be expected to be coordinated in the operation area, or the operations would be conducted near an airfield.

In the European Union, the current regulatory environment was recognized to be in such a state that the regulations allow complex operations through sufficient risk assessments and preparatory work from the UAV operator. For the scaled-up Logistics operations to be viable in European Union, the standardization community and the European Commission would have to produce new Standard Scenarios and harmonized standards for C-marked UAVs to allow a more streamlined process for the UAV logistic operators.

Meanwhile, before the harmonization and regulation processes proceed to producing the common rules for the STS operations, the standardization efforts conducted by, for example, CEN and ISO do allow the UAV manufacturers and operators to start leveling the play field by improving safety and interoperability between the UAVs, systems, parts, and operating procedures.

These findings were facilitated by applying an established CPS framework for an autonomous unmanned aerial system. Further research is needed to fully expand the analysis for the subsystems of an autonomous UAV system and get more detailed insight into the holes in current regulations or standards.

References

1. Hildmann, H., & Kovacs, E. (2019). Review: Using unmanned aerial vehicles (UAVs) as Mobile sensing platforms (MSPs) for disaster response, civil security and public safety. *Drones, 3*, 3.
2. Dow Jones Institutional News. (2015). *Pilots Are spotting more drones, prompting worries about collisions*. New York.
3. McIntyre, D. A. (2015). *Dozens of near midair collisions between drones and planes*. 24/7 Wall st.
4. Bryant, C., Foy, H., & Hollinger, P. (2015). Lufthansa jet in near-collision with drone: Airlines. In *The financial times* (London ed.).
5. Hamilton, G. (2017). Drone's collision with jet raises fears; no licensing requirements for operators. *Edmonton Journal*.
6. European Union Aviation Safety Agency (EASA). *"Drone Collision" Task Force*. https://www.easa.europa.eu/en/document-library/general-publications/drone-collision-task-force. Last accessed 28 May 2024.
7. European Union Aviation Safety Agency (EASA). *European Plan for Aviation Safety (EPAS) 2018–2022 Leaflet*. https://www.easa.europa.eu/en/document-library/general-publications/european-plan-aviation-safety-epas-2018-2022-leaflet. Last accessed 28 May 2024.
8. European Union Aviation Safety Agency (EASA). (2021). *Easy Access Rules for Unmanned Aircraft Systems (Regulations (EU) 2019/947 and (EU) 2019/945)*.

9. European Union. *Commission Implementing Regulation (EU) 2019/947 of 24 May 2019 on the rules and procedures for the operation of unmanned aircraft.* https://eur-lex.europa.eu/legal-content/EN/TXT/?uri=CELEX\%3A32019R0947. Last accessed 28 May 2024.
10. Nawaz, S. A. (2024). Regulating autonomy in civilian drones: Towards a spectral approach. *Journal of Intelligent & Robotic Systems, 110*(2), 46. ISSN: 0921-0296.
11. ICAO. (2018). *Annex 11—Air traffic services to the convention on international civil aviation.* ICAO.
12. Müller, R. (2021). Authorities and regulations in aviation. In R. Müller, A. Wittmer, & T. Bieger (Eds.), *Aviation systems* (Classroom companion: Business) (pp. 303–333). Springer.
13. Mohsan, S. A. H., Khan, M. A., Noor, F., Ullah, I., & Alsharif, M. H. (2022). Towards the unmanned aerial vehicles (UAVs): A comprehensive review. *Drones.*
14. European Commission (EC). (2021). Commission implementing regulation (EU) 2021/664 of 22 April 2021 on a regulatory framework for the U-Space. *Official Journal of the European Union 139.*
15. SESAR Joint Undertaking. (2023). *U-space concept of operations (CONOPS)* (4th edn.). SESAR Joint Undertaking.
16. European Union Aviation Safety Agency (EASA). (2024). *Easy Access Rules for U-space (Regulation (EU) 2021/664).*
17. Yi, H., Jolfaei, A., & Zheng, X. (2022). *Cyber-physical Systems in Transportation.* Springer.
18. Nardelli, P. H. J. (2022). *Cyber-physical systems.*
19. European Aviation Safety Agency. *EASA's regulatory role: increasing safety, environmental protection and enabling innovation via rulemaking.* https://www.easa.europa.eu/en/light/topics/easas-regulatory-role-increasing-safety-environmental-protection-and-enabling
20. European Union. *Regulation (EU) 2018/1139 on common rules in the field of civil aviation and establishing a European Union Aviation Safety Agency.* https://eur-lex.europa.eu/legal-content/EN/TXT/?uri=CELEX:32018R1139. Last accessed 18 November 2024.
21. European Union. *Commission Delegated Regulation (EU) 2019/945 of 12 March 2019 on unmanned aircraft systems and on third-country operators of unmanned aircraft systems.* https://eur-lex.europa.eu/legal-content/EN/TXT/?uri=CELEX\%3A32019R0945. Last accessed 28 May 2024.
22. European Union. *Regulation (EU) No 1025/2012 of the European Parliament and of the Council of 25 October 2012 on European standardisation.* https://eur-lex.europa.eu/legal-content/EN/TXT/?uri=CELEX:32012R1025
23. ISO. *Standards by ISO/TC 20/SC 16 Uncrewed aircraft system.* https://www.iso.org/committee/5336224/x/catalogue/p/0/u/1/w/0/d/0. Last ccessed 21 November 2024.
24. European Commission. *Harmonised Standards, Unmanned aircraft systems.* https://single-market-economy.ec.europa.eu/single-market/european-standards/harmonised-standards/unmanned-aircraft-systems. Last accessed 20 Novermber 2024.
25. Shakhatreh, H., et al. (2019). Unmanned aerial vehicles (UAVs): A survey on civil applications and key research challenges. *IEEE Access, 7.* https://doi.org/10.1109/ACCESS.2019.2909530
26. Adoni, W. Y. H., Lorenz, S., Fareedh, J. S., Gloaguen, R., & Bussmann, M. (2023). Investigation of autonomous multi-UAV systems for target detection in distributed environment: Current developments and open challenges. *Drones.* https://doi.org/10.3390/drones7040263
27. Elfatih, N. M., Ali, E. S., & Saeed, R. A. (2023). Navigation and trajectory planning techniques for unmanned aerial vehicles swarm. In *Artificial intelligence for robotics and autonomous systems applications* (Studies in computational intelligence). Springer.
28. Dinelli, C., Racette, J., Escarcega, M., Lotero, S., Gordon, J., Montoya, J., Dunaway, C., Androulakis, V., Khaniani, H., Shao, S., et al. (2023). Configurations and applications of multi-agent hybrid drone/unmanned ground vehicle for underground environments: A review. *Drones.* https://doi.org/10.3390/drones7020136

29. Utkarsh, A. (2023). UAV-based terrain-following mapping using LiDAR in high undulating catastrophic areas. In *Proceedings of UASG 2021: Wings 4 sustainability* (Vol. 304). Springer.
30. MAR, E., & Ndoma, A. (2019). The uses of unmanned aerial vehicles –UAV'S- (or drones) in social logistic: Natural disasters response and humanitarian relief aid. *Procedia Computer Science, 149.* https://doi.org/10.1016/j.procs.2019.01.151
31. Zhou, Y.-K., Rao, B., & Wang, W. (2020). UAV swarm intelligence: Recent advances and future trends. *IEEE Access, 8.*
32. Tahir, A., et al. (2019). Swarms of unmanned aerial vehicles. *Journal of Industrial Information Integration, 16.*
33. Gugan, G., & Haque, A. (2023). Path planning for autonomous drones: Challenges and future directions. *Drones.* https://doi.org/10.3390/drones7030169
34. Saffre, F., Karvonen, H., & Hildmann, H. (2024). Wild swarms: Autonomous drones for environmental monitoring and protection. In *New developments and environmental applications of drones.* Springer.
35. Rui, D., et al. (2023). UAV path planning based on deep reinforcement learning. In *Artificial intelligence for robotics and autonomous systems applications.* Springer.
36. Mukhamediev, R. I., Symagulov, A., Kuchin, Y., Zaitseva, E., Bekbotayeva, A., Yakunin, K., Assanov, I., Levashenko, V., Popova, Y., Akzhalova, A., et al. (2021). Review of some applications of unmanned aerial vehicles technology in the resource-rich Country. *Applied Sciences.* https://doi.org/10.3390/app112110171
37. Raman, R., et al. (2023). The role of 'unmanned aerial vehicles' in Smart City planning and management. In *Proceedings of UASG 2021: Wings 4 sustainability.* Springer.
38. Kotlinski, M., & Calkowska, J. K. (2022). U-space and UTM deployment as an opportunity for more complex UAV operations including UAV medical transport. *Journal of Intelligent & Robotic Systems, 106*, 1.
39. Li, F., & Kunze, O. (2023). A comparative review of air drones (UAVs) and delivery bots (SUGVs) for automated last mile home delivery. *Logistics.* https://doi.org/10.3390/logistics7020021
40. Elkhapery, B., & Small, E. P. (2022). Exploratory research towards automated masonry construction using UAVs. *IOP conference series Materials Science and Engineering.*
41. Singh, C. H., Mishra, V., & Jain, K. (2023). Review of Uncrewed aerial vehicle swarm system coordination and communication. In *Proceedings of UASG 2021: Wings 4 sustainability* (Lecture notes in civil engineering). Springer.
42. Darush, Z., et al. (2023). SwarmGear: Heterogeneous swarm of drones with morphogenetic leader drone and virtual impedance links for multi-agent inspection. In *2023 International conference on unmanned aircraft systems (ICUAS).* IEEE.
43. Jain, K., Mishra, V., & Pradhan, B. (2023). Comparison of DEM generated from UAV images and ICESat-1 elevation datasets with an assessment of the cartographic potential of UAV-based sensor datasets. In *Proceedings of UASG 2021: Wings 4 sustainability.* Springer.
44. Weng, K. C., Lin, S. T., Hu, C. C., et al. (2023). Multi-view approach for drone light show. *The Visual Computer, 39*, 5797–5808. https://doi.org/10.1007/s00371-022-02696-8
45. Svanström, F., Alonso-Fernandez, F., & Englund, C. (2021). A dataset for multi-sensor drone detection. *Data in Brief, 39.* https://doi.org/10.1016/J.DIB.2021.107521
46. NATO. (1995). *STANAG 4347 Definition of nominal static range performance for thermal imaging systems.*
47. MAVLink Developer Guide. https://mavlink.io/en/
48. Karvonen, H., et al. (2023). Using a semi-autonomous drone swarm to support wildfire management—A concept of operations development study. In *Engineering psychology and cognitive ergonomics* (Lecture notes in computer science). Springer.
49. Finnish Transport and Communications Agency. *OPS M1–35: Valtion miehittämätön ilmailu.* https://finlex.fi/fi/viranomaiset/normi/498001/47638

50. Dauer, J. C., et al. (2021). Automated cargo delivery in low altitudes: Concepts and research questions of an operational-risk-based approach. In *Automated low-altitude air delivery*. Springer.
51. Nikodem, F., et al. (2021). Operations risk based concept for specific cargo drone operation in low altitudes. In *Automated low-altitude air delivery*. Springer.

Standardizing Drone Remote Sensing Data for Agricultural Data Space Ecosystems

Petri Linna[1(✉)], Antti Halla[1], and Juha Suomalainen[2]

[1] Tampere University, Tampere, Finland
petri.linna@tuni.fi
[2] Finnish Geospatial Research Institute of National Land Survey of Finland, Espoo, Finland
https://www.tuni.fi

Abstract. The integration of multispectral drone data from Drone-in-a-Box (DiaB) systems into data spaces for agriculture is a complex task. It requires standardized methods and supporting infrastructure. The diverse nature of multispectral data, along with the vast volume collected and the technical expertise required for its processing, presents significant challenges, particularly for agricultural applications where precise data is crucial. This research began with a literature review to identify relevant data standards and their potential applications in processing multispectral drone data. Following this, experimental field tests were conducted, where imagery was collected using a DiaB system. The data were processed using the OpenDroneMap application to explore how it can be transformed into standardized formats suitable for integration into agricultural data spaces. The objective of this study is to explore how drone-based multispectral data, when processed through open-source tools, can be made compatible with agricultural data space ecosystems. The study focuses on the roles of metadata, standards (such as INSPIRE, DCAT, and ISO standard family), vocabularies and terms (such as Agrovoc and Dublin Core), API technologies, and data formats that support structured data exchange in agriculture. The results provide practical insights and technical recommendations for making drone data more accessible and reusable. A prototype SPARQL validation query was also developed to align metadata with agricultural vocabularies. These findings support the broader vision of enabling data-driven agriculture through standardized and interoperable data infrastructure. Future research should explore how this data can be integrated into agricultural data space marketplaces and evaluate its commercial potential. Additionally, there is a need to explore how smart contracts can be implemented for automated drone data management, further facilitating seamless and fair data transactions in data space marketplaces.

Keywords: Drone · standards · Data · OpenDroneMap · Data space · Metadata · DCAT

M. Laakkonen (ed.), *DroneAI for Resilient Nature*,
https://doi.org/10.1007/978-3-032-19008-6_11

1 Introduction

1.1 Background

The upcoming Common Agricultural Policy (CAP) incorporates a forward-looking vision where the use of data in agriculture is expected to bring significant benefits. The strategy promotes a shift towards data-driven management, allowing a more flexible and real-time approach to agricultural policy. Instead of fixed calendar-based actions, the focus is on verifying performance through measurable indicators. These may include environmental impact, resource efficiency, and crop productivity [1].

Drones have become essential tools in modern precision agriculture, enabling farmers to collect real-time data on crop health, soil conditions, and field variability [2–5]. Their ability to cover large areas quickly while capturing detailed multispectral and thermal imagery provides a clear advantage over traditional monitoring methods. Several studies [4–6] highlight how Unmanned Aerial Vehicles (UAVs) are used for crop monitoring, yield prediction, pest detection, irrigation planning, and disease monitoring, all of which support more efficient and informed decision-making.

The complexity and volume of multispectral data, combined with the technical expertise required for processing, create significant barriers. In particular, integrating drone data into farm management systems or broader data ecosystems remains a challenge. Agricultural data comes from various sources, including satellite imagery, field sensors, machinery, and Internet of Things (IoT) devices. This diversity makes it essential to harmonize how data is collected, processed, and shared [4,5].

Recent literature emphasizes the need for standardizing data models and processing workflows to make drone data more useful on a larger scale [7,8]. The Food and Agriculture Organization (FAO) also stresses the importance of improving data quality and reuse in agriculture. This includes clear data-sharing principles and suitable licensing frameworks to support responsible and effective data exchange [9].

Looking ahead, drones are expected to incorporate more onboard intelligence, including real-time AI-based analysis during flight. The geospatial software market related to drone applications is also projected to grow at over 16% annually, underlining the increasing relevance of scalable and standardized drone data workflows [4].

These findings underline the relevance of this study, which investigates how drone data can be processed and aligned with standardized metadata models to improve interoperability across agricultural systems.

1.2 Problem Statement

One of the main barriers to effective drone use in agriculture is the significant expertise required throughout the entire workflow. Preparing the drone and planning the image capture process are time-consuming tasks. The following image processing phase is often even more demanding. Only after these steps can the analysis begin, which requires interpreting the data correctly and often combining it with other datasets. Additional skills are needed if the results are to be used in agricultural machinery or external systems.

To reduce costs and improve scalability, it would be beneficial to automate the full chain of drone data collection, post-processing, and delivery. This would also open possibilities for reusing the data in research or commercial services. For example, drone imagery could be sold to third parties or used in decision support systems. However, implementing such automation raises new challenges. It is unclear which tools or platforms are suitable, how data should be structured, and what quality requirements must be met.

This study focuses on finding solutions to these challenges. It explores how a complete drone data pipeline can be designed, from image acquisition to standardized output, in a way that enables integration into agricultural data spaces. In this context, the term "*agricultural data space*" refers to a trusted digital environment where agricultural data can be shared, accessed, and reused among stakeholders using common standards. The aim is to improve compatibility, support reuse, and lay the foundation for potential commercialization.

1.3 Objectives and Research Goals

The main objective of this study is to explore how multispectral camera data collected by drones, processed using the OpenDroneMap (ODM) application, can be transformed into a standardized format compatible with agricultural data spaces. The research focuses on key components that enable this transformation, including metadata standards, controlled vocabularies, data formats, and API technologies.

The study examines how standards such as INSPIRE, DCAT, and the ISO 19115 family, as well as vocabularies like AGROVOC and Dublin Core, support metadata harmonization. It also considers the role of API interfaces such as REST and Swagger in enabling data exchange between systems. The processing pipeline is evaluated using open-source tools, with a focus on producing outputs that are ready for integration into larger data infrastructures.

The broader goal is to support the wider adoption of drone-based monitoring in agriculture by improving data interoperability and reducing the technical barriers to reuse. The study contributes to ongoing efforts to make agricultural data more accessible, semantically aligned, and commercially viable in future digital ecosystems.

1.4 Paper Structure

This paper is structured as follows. Chap. 2 provides a literature review of relevant standards, vocabularies, and technical frameworks that support the integration of drone data into agricultural data spaces. Chapter 3 presents the research setup, including the drone docking system used in the field tests, the data processing workflow based on open-source tools, and the approaches applied to metadata evaluation. Chapter 4 summarizes the main findings, such as the compatibility between ODM outputs and metadata standards, as well as the development of a prototype query for vocabulary validation. Chapter 5 discusses the broader implications of the results, particularly in relation to standardization, data reuse, and commercialization potential. Finally, Chap. 6 concludes the paper and outlines directions for future research.

2 Literature

Standardizing drone data for agricultural applications requires an understanding of existing frameworks, tools, and practices that support interoperability and reuse. This chapter reviews relevant literature on data standards, metadata models, and technical approaches that enable the integration of drone-generated imagery into agricultural data spaces. The focus is on identifying established solutions that help structure, describe, and share drone data in a consistent and machine-readable way.

Special attention is given to metadata standards such as ISO 19115 and DCAT, domain-specific vocabularies like AGROVOC and Dublin Core, and infrastructure models that support trusted and scalable data exchange. These elements form the foundation for developing workflows that make drone data compatible with broader agricultural information systems.

2.1 Relevant Standards

This study began with a literature review to identify data standards and terminology relevant to the processing and integration of drone imagery data. The perspective was then expanded to explore the utilization of data spaces, with a focus on key considerations for producing and distributing data within a trusted environment. Such spaces not only enable secure and interoperable data sharing but also allow new approaches to data licensing and exchange, which differ from more traditional data-sharing models.

ISO 19115 (Geographic Information—Metadata) is one of the most commonly referenced standards for geographic metadata. It standardizes how digital geospatial data is described, helping define metadata requirements for data discovery, evaluation, and use [10].

INSPIRE, developed under the European Commission, is designed to improve spatial data sharing across Europe [11]. It is particularly relevant in agriculture for managing datasets such as field boundaries and ensuring interoperability between geographic datasets.

Sensor Model Language (SensorML) is part of the Open Geospatial Consortium (OGC) standards, and it provides an XML-based framework for encoding sensor data, including drones [12]. This standard can be valuable when integrating real-time sensor observations into agricultural datasets.

ISO maintains several technical committees dedicated to developing standards. The ISO/TC 20/SC 16 committee focuses on international standards for safe and efficient use of UAS globally [13]. CEN/TC 471/WG1 develops UAS-related standards for the European Union, supporting regulatory and safety needs in alignment with EU legislation. There are a few significant ISO standards related to UAVs.: ISO 21384-1 is a comprehensive standard that covers the general requirements for unmanned aircraft systems. It addresses aspects related to the safety, performance, and operation of UAS, but it also includes requirements for data management and reliability. The standard ensures that data collected by UAS can be processed and shared securely and efficiently. ISO 21384-3 standard focuses on the operation of UAS, including data collection and sharing in various use scenarios. It specifically addresses not only operational and regulatory requirements but also standards related to data processing.

In addition to formal standards, FAIR data principles, are widely applied in agriculture. These principles emphasize making data findable, accessible, interoperable, and reusable. These principles are increasingly being applied to ensure the reusability of data across various platforms, including data spaces [14].

This study also uses ODM [15] in its experimental work. ODM is an open-source software tool for processing drone imagery into standardized file formats such as GeoTIFF, making the data compatible with other geographic information systems (GIS). This tool was selected for its ability to structure drone imagery in a way that supports downstream integration.

Finally, this paper reviews some standards that are foundational to data space design. These standards highlight the importance of interoperability and metadata completeness. In agricultural data spaces, domain-specific metadata is essential for enabling accurate analysis and data reuse.

DCAT (Data Catalog Vocabulary) is a W3C standard, which is used for publishing data catalogs on the web [16]. It plays a central role in making datasets discoverable and is often used in conjunction with agricultural data spaces.

The Data Space Support Centre (DSSC), as part of the broader European Data Strategy, provides essential resources and guidance for the creation and implementation of data spaces across different domains. The Blueprint v1.5, a key document published by the DSSC, serves as a comprehensive guide for designing, implementing, and managing interoperable data spaces. It highlights the importance of using established standards such as DCAT v3 and Open Digital Rights Language (ODRL) and Dataspace Protocol [17,18].

The Data Space Protocol is a key component of the International Data Spaces (IDS) architecture, providing a standardized framework for secure and trusted data exchange between participants in a data space. It defines the rules, interfaces, and interactions necessary for establishing data sharing agreements, managing data access, and facilitating the exchange of data while preserving data sovereignty. The protocol emphasizes the use of DCAT to ensure successful data transfer and interoperability. By leveraging DCAT, the protocol standardizes metadata descriptions, enhancing discoverability and accessibility of datasets across different platforms [19].

2.2 Vocabulary and Terms

Standards and terminologies often overlap significantly. The following section describes two examples that are more vocabulary-based rather than full standards, although they are also prominently featured within various standards.

The Dublin Core Metadata Element Set is a widely used vocabulary for describing resources such as data sets, documents, or images. It provides a core set of metadata elements including for example title, creator, subject, and date. It is flexible and can be applied to various types of digital resources, making it particularly useful in data spaces where multiple types of data are integrated. In agricultural data spaces, it helps ensure that drone data is described in a consistent and machine-readable way. Dublin Core is often used in conjunction with other metadata standards like ISO 19115 and DCAT, increasing its compatibility across systems [20].

AGROVOC is a multilingual vocabulary managed by FAO. It includes terms related to agriculture, fisheries, forestry, and related domains. It is specifically designed to facilitate the sharing and retrieval of agricultural knowledge and data [21]. AGROVOC can be used to tag and categorize drone data relevant to agriculture. Using AGROVOC ensures that all data participants (farmers, agronomists, researchers) have a consistent understanding of the terms used to describe the data. It supports data interoperability and reuse, as everyone in the ecosystem can rely on the same vocabulary. This makes AGROVOC especially useful in cross-border or multilingual agricultural data spaces.

Using controlled vocabularies like Dublin Core and AGROVOC helps improve data discoverability, alignment, and reuse. These vocabularies provide a foundation for integrating drone data into broader agricultural information systems in a structured and interoperable way.

2.3 From Data Platforms Towards Data Spaces

Data spaces represent a significant evolution in data management, enabling secure and interoperable data sharing across sectors while maintaining data sovereignty. Unlike traditional isolated platforms, data spaces promote secure and interoperable data exchange based on common rules and governance models. This concept, championed by the International Data Spaces Association (IDSA), emphasizes interoperability, privacy, and control, addressing the need for decentralized and interconnected data environments [22].

Agri-Gaia is a notable example of a data space initiative for agriculture sector. It is part of the broader Gaia-X [23,24] initiative, which aims to create a European cloud and data ecosystem that aligns with the principles of transparency, openness, and interoperability. Agri-Gaia seeks to provide a platform for farmers, technology providers, and other stakeholders in the agricultural value chain to share data securely and collaboratively, promoting innovation in smart farming through AI and machine learning [24,25].

While Agri-Gaia focuses on creating a secure platform for agricultural data sharing, organizations like Big Data Value Association (BDVA) provide technical and ethical guidelines to ensure broader interoperability across sectors. As a key partner in the European Data Spaces initiative, BDVA focuses on creating frameworks for data interoperability and ensuring that big data technologies align with data sovereignty principles. BDVA's contributions include defining technical and ethical guidelines for data sharing within data spaces, with an emphasis on fostering innovation across industries like agriculture, healthcare, and manufacturing [26].

Data spaces are designed to overcome fragmentation. Currently, many agricultural data platforms operate in isolation, forming silos that limit the broader reuse of valuable data [27,28]. In addition, there are numerous standards that must be considered when designing and building comprehensive data space ecosystems [29].

Data spaces provide a solution by ensuring that data can flow securely between different stakeholders (e.g., farmers, researchers, agribusinesses), allowing for more efficient decision-making and innovation. This collaborative approach is essential in agriculture, where data-driven insights from weather conditions, crop health, and soil monitoring can significantly improve productivity and sustainability.

FIWARE is an open-source platform that supports the development of smart applications in multiple sectors, including agriculture. It plays a key role in enabling the creation of smart farming applications by providing data models, APIs, and context information management tools that facilitate the integration of IoT, drones, and sensor data into agricultural data spaces. FIWARE is already widely used in smart city and industrial data spaces and holds potential for integrating multispectral drone data into agricultural data ecosystems [30,31].

The AgriDataSpace report highlights several implementations of data platforms in agriculture, but none explicitly mention drones. This reflects the current fragmentation of technologies and platforms in the agricultural sector. Multiple technological implementations exist for managing farm data, but they often lack integration with more advanced data sources like drones. This fragmentation creates challenges for scaling and adopting data-driven innovations in agriculture, where data must be aggregated from multiple sources, including IoT sensors, farm machinery, and drones [27].

To address these challenges, the Simpl initiative has been introduced. Simpl aims to create a unified framework for developing and governing data spaces, ensuring that different data sources, including drones, can be seamlessly integrated into a standardized and interoperable ecosystem. This initiative is particularly important for agriculture, where fragmented solutions currently hinder the full potential of data-driven farming. By providing guidelines for structuring and managing data, Simpl can help align agricultural data spaces towards a more cohesive and scalable model. The work on Simpl began in 2024 [27,32].

The Data Spaces Business Alliance (DSBA) is an initiative by four major European organizations: Gaia-X, IDSA, BDVA, and FIWARE Foundation. It promotes collaboration around data spaces [33].

The alliance was created to drive the development and adoption of data spaces across Europe, fostering secure, interoperable, and sovereign data sharing ecosystems across industries. The DSBA promotes collaboration and sets the groundwork for innovation by creating standardized guidelines and frameworks that ensure data interoperability and trust. The focus is on establishing data ecosystems in various sectors, such as healthcare, energy, agriculture, and manufacturing. The Alliance leverages the expertise of its founding members to ensure that the infrastructure, data standards, and policies are aligned across sectors, promoting data sovereignty while encouraging innovation and collaboration across borders.

In addition, the Agricultural Information Management Standards (AIMS) initiative, led by FAO, provides a comprehensive framework of standards, tools, and methodologies to enhance the management, interoperability, and accessibility of agricultural information. AIMS promotes the adoption of FAIR data principles and supports global agricultural data sharing through controlled vocabularies, metadata standards for vocabulary management [34].

According to Wolfert [35], the future of smart farming could evolve in two directions: closed proprietary systems or open collaborative ecosystems. The latter approach provides stakeholders with greater flexibility in selecting business partners and technological solutions, potentially reshaping roles and power dynamics across the value chain. The emergence of data spaces and the implementation of EU regulations (for example Data Act [36]) create opportunities to safeguard stakeholders who may have less negotiating power.

As highlighted in this section, data spaces are a broad concept with numerous stakeholders involved. New data spaces initiatives [27,28,37] are continually emerging making it crucial to ensure their managed growth and interoperability. Different data spaces must be capable of facilitating seamless data exchange among themselves, as businesses and other entities will likely operate across multiple data spaces in the future.

From the perspective of drone data, it is unlikely that a dedicated drone-specific data space will be established. Instead, drone data is expected to be integrated into multiple data spaces. For this reason, the standardization of drone data becomes even more critical. Rather than developing tailored solutions for a single data space, it is essential to adopt standards that support interoperability across multiple data spaces, ensuring broader applicability and usability, but also more business possibilities. This enables versatile data reuse and broader utilization of GIS applications.

Despite the rapid growth in drone technology, the integration of drone data into agricultural data spaces remains limited. Drone data, especially multispectral and thermal imagery, is crucial for precision farming, but it often remains siloed due to lack of standardized frameworks for its integration into larger data ecosystems. Standardizing drone data formats, metadata, and ensuring its compatibility with data space technologies like DCAT are essential next steps for enhancing data interoperability in agriculture.

3 Research

The previous chapter provided an overview of relevant standards, vocabularies, and the broader context of data spaces. This chapter presents the applied part of the study, focusing on the implementation and evaluation of a drone data processing workflow. The chapter begins by describing the Drone in a Box (DiaB) system used in the field tests, including its technical components and operating principles. It then outlines the data processing framework based on ODM, followed by an analysis of metadata compatibility with existing standards. The chapter also covers the use of controlled vocabularies, prototype development for metadata validation, and the role of API technologies in data delivery. Finally, it examines the potential for commercializing standardized drone data in data space environments.

Figure 1 illustrates the drone data collection and processing workflows. Automatic drone flights are operated using the DiaB system. The next block in the figure represents the framework for data processing, including the preprocessing of drone data. This is followed by a description of API interfaces, enabling connections to either data space marketplaces or GIS applications. This process is also described in detail in [39].

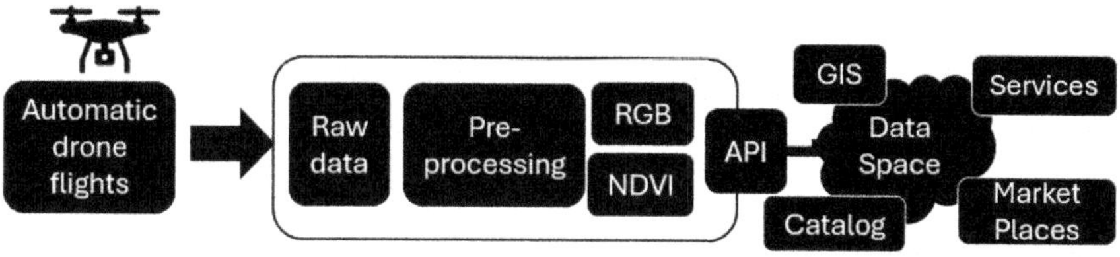

Fig. 1. Drone data collection process (modified from [39])

3.1 DiaB Systems

A DiaB system consists of a drone and a docking station which offers automated services such as recharging of batteries and offloading data. These stations are typically connected to the power network or equipped with solar panels, allowing drones to autonomously land, recharge, and resume their missions without manual intervention. A docking station enhances automation in drone operations, enables drones to be kept on standby, and facilitates repeated operations at a single location. Depending on the legislative restrictions, they may also allow fully remote operation of the drones without a pilot on site.

DiaB systems enhance the efficiency of drones in remote or large-area applications like agriculture, environmental monitoring, or surveillance by extending operational range and uptime. These stations can offload data collected by the drone for further processing, facilitating near-real-time analytics.

In this study, the DiaB system used was a *Rumble Tools HK-drone* (Rumble Tools, Imatra, Finland) mounted with *Micasense Altum-PT* multispectral camera (AgEagle, Wichita, Kansas, USA) and *Applanix APX-15 EI UAV* GNSS-IMU system (Trimble Applanix, Richmond Hill, Ontario, Canada). This configuration enabled high-resolution image capture and precise geolocation of the collected data. The drone was deployed in test flights during the summer of 2024 to collect multispectral imagery for further processing and analysis, with the primary focus placed on evaluating the data workflows themselves (Fig. 2).

Fig. 2. Rumble tools HK-drone used in test flights

3.2 Framework of Drone Data Processing

Data collected from drones must be preprocessed before it can be effectively used. This typically involves stitching images into orthophotos, applying coordinate corrections, adjusting color, and performing other refinements. After preprocessing, various data products can be generated based on specific needs, such as RGB or NDVI maps.

The following section describes some of the most commonly used commercial and open-source solutions that enable versatile management of drone data. It is also important to note that after preprocessing, the imagery can often be further analyzed using GIS applications such as ArcGIS and QGIS.

DJI is one of the leading commercial players in the drone industry. Its platforms, such as DJI FlightHub, support efficient data management and sharing within organizations. DJI also offers photogrammetry software like DJI Terra, which enables users to process and visualize drone data in real time. These tools make the data accessible to multiple stakeholders and support collaboration across teams [40].

Table 1. Key output files generated by OpenDroneMap

File name	Description
0-0-0-0.json	Stores information related to the tile index or geospatial data split for processing. It includes metadata about the specific area of the project that has been processed or georeferenced
cameras.json	Contains metadata about the cameras used for capturing images, such as camera model, sensor size, focal length, and calibration details. It helps in understanding the properties of the imagery for accurate processing
ept.json	Part of the Entwine Point Tile (EPT) format, used for organizing and storing large point clouds. It provides the structure and metadata for point cloud data, aiding in efficient rendering and querying
ept-build.json	Stores information about the build process for the EPT dataset, including processing parameters, version, and configuration used during the creation of point cloud tiles
images.json	Contains a list of all images used in the project, including metadata such as file paths, GPS coordinates, and orientation. This file links the raw imagery data to the processed outputs
log.json	Records the processing log and execution details of the OpenDroneMap workflow. It includes information about each processing step, errors encountered, and system performance metrics
manifest.json	Provides a summary of the entire project and its output files. It lists the generated files, their locations, and metadata about the processing run, useful for quick reference and tracking project completeness
odm_georeferenced_model.boundary.json	Contains geospatial boundary information of the georeferenced model, outlining the extent of the model area in geographic coordinates, useful for defining the area of interest
odm_georeferenced_model.info.json	Provides detailed information about the georeferenced 3D model, including metadata like the coordinate system, scaling factors, and other geospatial properties critical for accurate alignment
odm_georeferenced_model.json	Represents the main georeferenced 3D model data, typically including vertex coordinates, faces, and other elements needed for rendering the model in visualizations and GIS applications
odm_georeferenced_model.summary.json	Offers a concise summary of the georeferenced model, including basic statistics and metadata such as the number of vertices, faces, and overall file size, useful for understanding the complexity of the model
stats.json	Provides statistical information about the processing run, such as processing time, resource usage (CPU, memory), and data quality metrics, useful for performance analysis and debugging

Pix4D offers several solutions for sharing and distributing drone data through its Pix4Dcloud platform. This cloud-based service enables users to upload drone-captured imagery and generate 2D maps and 3D models, which can then be shared with stakeholders. Pix4D also supports integrations with platforms such as Trimble Connect to enhance team collaboration. In addition, Pix4Dengine provides an API that allows businesses to embed Pix4D's photogrammetry capabilities into their own infrastructure, enabling scalable drone data processing and sharing [41].

Optelos offers a comprehensive platform that integrates drone data into digital twin environments and geospatial systems. It supports visual data management by enabling secure storage, sharing, and analysis of drone-collected imagery [42].

OpenAerialMap is an open-source initiative that aims to create a global repository of openly licensed aerial imagery. Although it does not process imagery directly like ODM, it offers a platform for sharing, storing, and accessing aerial data, making it a valuable complement to other drone data processing tools [43].

ODM is an open-source software ecosystem for processing aerial imagery captured by drones into geospatial data. It converts raw imagery into 3D models, orthophotos, point clouds, and other formats compatible with GIS software for analysis and visualization. The software is widely used in agriculture, construction, environmental monitoring, and other domains that rely on spatial data [15].

WebODM is a fully open-source, web-based interface for ODM. It provides a user-friendly environment for processing drone imagery and produces outputs such as orthophotos, point clouds, and 3D models using the same open-source processing pipelines [44].

As the list above illustrates, various software solutions are available for processing drone data. This paper does not compare their performance but instead focuses on a specific example by analyzing the dataset generated using ODM.

In our use case, ODM produces a set of JSON files that describe the data from multiple perspectives. Table 1 lists these files along with their respective descriptions. ODM also creates a directory structure to organize different output datasets. Some of the generated files contain extensive metadata and contextual information.

3.3 Integration of Standards and Vocabularies

This section examines how metadata standards and vocabularies can be compared to identify shared elements and support interoperability. DCAT was selected for analysis because it has been widely adopted as the standard for data spaces [17]. INSPIRE was included due to its strong relevance to geospatial data, and Dublin Core was used to represent general-purpose metadata elements. ISO 19115 was excluded from the comparison, as INSPIRE builds upon it directly.

The comparison is presented in a metadata mapping table (Table 2), which shows how common fields are expressed across these standards. The left column lists metadata terms from the 15 core elements of Dublin Core [20], followed by a short description. The next columns show equivalent or closely related elements in DCAT, Dublin Core Terms, and INSPIRE. Where no direct equivalent exists, the corresponding cell is left empty or marked as "object" to indicate partial similarity.

It is important to note that DCAT incorporates extended Dublin Core elements directly. For example, *dc:title* refers to the basic Dublin Core element, whereas *dct:title* refers to the extended version. These distinctions are important when aligning metadata structures across different systems.

Abbreviations used in the table: dc = Dublin Core (basic), dct = Dublin Core Terms (extended), dcat = DCAT vocabulary, gmd = INSPIRE metadata elements. The abbreviations for the standards have been sourced from publicly available resources.

The comparison revealed that the analyzed standards are largely similar in structure and intent. This supports the idea that metadata from drone data workflows can be mapped and transformed with relatively modest effort. Although the table only covers a small subset of available fields. Dublin Core

Table 2. Metadata comparison

Metadata element	Description	Dublin core	DCAT	INSPIRE	Schema.org
Title	The title or name of the resource	dc:title	dct:title	gmd:title	sdo:name
Description	A brief description of the resource content	dc:description	dct:description	gmd:description	sdo:description
Creator	The person or organization responsible for creating the resource	dc:creator	dct:creator		
Publisher	The entity responsible for publishing the resource	dc:publisher	dct:publisher		sdo:publisher
Contributor	An entity that contributed to the creation of the resource	dc:contributor	dct:contributor		
Date	The date related to the resource (e.g., publication date)	dc:date	dct:issued, dct:modified	gmd:date (gmd:dateStamp, gmd:CI_Date)	sdo:datePublished, sdo:dateModified
Identifier	A unique identifier for the resource	dc:identifier	dct:identifier	gmd:Identifier (gmd:fileIdentifier)	sdo:identifier
Subject	Topics describing the resource content	dc:subject	dct:subject		sdo:about
Type	The nature or genre of the resource	dc:type	dct:type		sdo:additionalType
Format	The physical or digital format of the resource (e.g., PDF, CSV)	dc:format	dct:format, dcat:mediaType	gmd:distributionFormat	sdo:encodingFormat, sdo:fileFormat
Source	The original resource from which the current resource is derived	dc:source	dct:source		
Language	The language of the resource content	dc:language	dct:language	gmd:language	sdo:inLanguage
Relation	A reference to a related resource	dc:relation	dct:relation		sdo:isRelatedTo
Coverage	The spatial or temporal extent of the resource	dc:coverage	dct:spatial, dct:temporal	(e.g. gmd:ex_geographicboundingbox, gmd:EX_BoundingPolygon, gmd:EX_GeographicDescription, gmd:EX_TemporalExtent, gmd:EX_SpatialTemporalExtent, gmd:MD_ReferenceSystem, gmd:EX_VerticalExtent, gmd:MD_Resolution)	sdo:spatialCoverage, sdo:temporalCoverage
Rights	Information about rights held over the resource	dc:rights	dct:rights, dct:license	gmd:accessConstraints, gmd:useConstraints	sdo:license

alone defines 55 elements, but this provides a useful starting point. The data space catalogs are expected to adopt DCAT, and this alignment facilitates interoperability across platforms. The effort needed to convert data into this standard is not as extensive as the number of available standards might initially suggest, especially when comparing the 15 core terms. This alignment simplifies the path toward creating interoperable and reusable data within data spaces.

3.4 Metadata in Drone Data

Barbieri et al. [45] conducted a survey among various stakeholders to identify the minimum requirements for drone datasets. The result was the Minimum Information Framework for Drone Datasets (MIF), which outlines the essential components needed for structured and reusable data.

Figure 3 presents an overview of the MIF structure. The dataset is divided into eleven sections. The "Project" section includes general information about the mission, such as its objectives and context. The "Individual Flight" section is further divided into "Platform & Payload," "Mission Plan," and "Flight Check and Calibrations," each describing different aspects of the flight configuration. The next section, "Data from Flight," includes the "Flight Log," "Individual

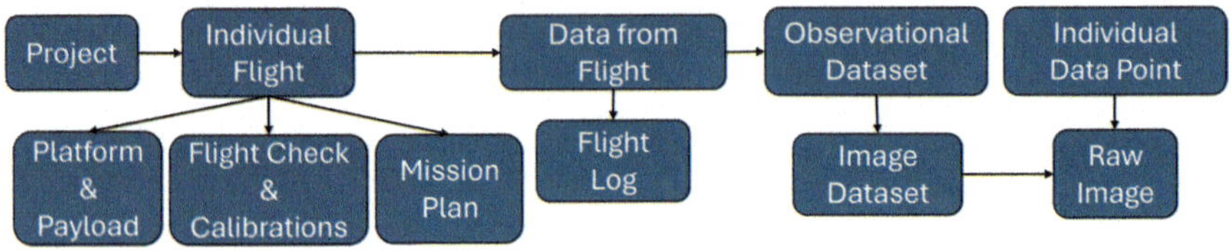

Fig. 3. Minimum information framework for drone datasets (modified from [45])

Data Point," and "Image Dataset." These parts focus on the actual data collected, including raw imagery and associated telemetry. The framework defines 74 individual attributes, categorized by importance based on stakeholder feedback. The four categories range from non-essential to absolutely critical for data reuse.

In this study, ODM was used to process the collected imagery. The software produces several JSON files, including *odm_georeferenced_model.json*, *cameras.json*, and *stats.json* . These files describe camera parameters, georeferencing details, and processing statistics. Some elements suggested in MIF, such as flight logs and calibrations, may be partially covered by *log.json*. However, a detailed attribute level comparison was not conducted.

Instead, the study used MIF as a conceptual reference to assess the completeness of the metadata produced by ODM. Based on this structure, it is possible to extract relevant fields and use them to build more comprehensive metadata packages. These could be dynamically generated depending on the context or target use case, whether for research, public distribution, or commercial marketplaces.

3.5 Tools for Aligning Drone Imagery and Terminologies

Processing drone imagery and aligning it with standardized vocabularies is a key step in ensuring interoperability and reuse. Drone data, especially multispectral imagery and 3D models, contains complex metadata that must be structured according to recognized standards to enable integration into agricultural data spaces.

Several tools and frameworks are available for this purpose. Validation tools such as the DCAT-AP validator [46] can check whether metadata descriptions meet the requirements of specific standards. JSON Schema can also be used to validate the structure of metadata, and Python libraries such as *jsonschema* support this process [47].

For vocabulary alignment, the Simple Knowledge Organization System (SKOS) provides a framework for mapping concepts across controlled vocabularies [48]. Tools like VocBench allow users to manage OWL ontologies and harmonize terms across datasets [49]. Apache Jena supports integration of metadata with RDF-based vocabularies, improving semantic consistency [50].

In terms of cataloging, *pycsw* is an open-source metadata catalog service that implements the OGC CSW standard [51]. GeoNetwork is another widely used tool that supports DCAT and INSPIRE, and enables the creation and validation of custom metadata schemas [52].

As part of this study, a prototype query was developed to compare terms generated by ODM with the AGROVOC vocabulary. The query extracted 357 terms from the processed data and queried AGROVOC using its SPARQL API. A total of 20 exact matches were found. While the match count was limited, this result was expected, as ODM terminology is not designed for agriculture-specific contexts. Nevertheless, the prototype provided a valuable test case for evaluating compatibility and establishing a baseline for future improvements.

In addition, we conducted early testing with GeoNetwork to build a metadata catalog. Although still in a preliminary phase, these experiments provided useful insights into how drone metadata can be structured, validated, and linked to standardized vocabularies. Together, these tools and experiences offer a foundation for developing workflows that support semantic alignment and enable more effective data integration.

3.6 Data Delivery Techniques

In the context of agricultural data spaces, a multitude of technical implementations exist for data delivery, as highlighted in the Agri Data Space report [27]. These solutions vary in complexity and capabilities, but all aim to support efficient and standardized data exchange across platforms.

Representational State Transfer (REST) is an architectural style rather than a specific technology. It leverages standard HTTP methods (GET, POST, PUT, DELETE) and typically uses JSON or XML for data exchange. RESTful APIs are widely used due to their simplicity, scalability, and ease of integration across different platforms.

GraphQL is a query language for APIs originally developed by Facebook. Unlike REST, GraphQL allows clients to request exactly the data they need, rather than receiving predefined structures. This flexibility helps reduce data over fetching and under fetching, making it particularly effective in applications with dynamic data needs [53].

Swagger (now part of the OpenAPI ecosystem) is an open-source framework that supports the design, documentation, and implementation of RESTful web services. The core component of Swagger is the OpenAPI Specification, a standardized, machine-readable document (often in YAML or JSON format) that defines the structure of the API. This includes details about available endpoints, HTTP methods (such as GET, POST, PUT), parameters, expected responses, and authentication methods. By providing a clear, standardized API definition, Swagger enhances interoperability and facilitates easier integration with various client applications [54].

These data delivery techniques are essential for enabling effective communication between data providers and consumers within agricultural data spaces. By adopting standardized technologies such as OpenAPI, REST, and GraphQL, data exchange becomes more interoperable, maintainable, and scalable across diverse platforms.

In this study, interface development is ongoing. Based on earlier experience and common practice in data space projects, Swagger has been selected as the preferred option for implementation. Before proceeding, it is necessary to define the data package formats to ensure compatibility with data space requirements. Once the structure is established, the API layer can be built accordingly. It should also be noted that integration into data spaces may require a dedicated data connector component, adding an additional layer to the interface implementation [39].

3.7 Data to Marketplace

The commercialization of drone data presents significant opportunities, but also introduces distinct challenges, particularly within the context of data spaces. As highlighted in recent reports [27,28], current implementations reveal a fragmented landscape of technologies and standards. Despite the rapid adoption of drones for data collection, a gap remains in fully integrating this data into operational data space marketplaces.

One promising approach involves the use of smart contracts, which can automate data sharing processes and enable monetization for data providers. These contracts enhance transparency and traceability, while also aligning data exchange practices with the FAIR principles [31].

Several challenges must be addressed to realize this potential. Metadata quality and standardization are key issues. Inconsistent metadata makes it difficult to index and reuse datasets. For drone data to be market-ready, it must be properly georeferenced, enriched with contextual information such as weather or soil data, and accompanied by standardized metadata descriptions based on formats like DCAT or INSPIRE.

Legal and regulatory considerations are also essential. Data sharing must comply with national and EU-level legislation concerning data ownership, privacy, and usage rights. Clear legal frameworks are needed to protect all participants, from smallholder farmers to technology providers.

Previous research, including the work by Barbieri et al. [45] , suggests that a single universal data model for drone data is unlikely to satisfy the diverse needs of all users. Different users have different requirements depending on their role. Researchers may need detailed calibration data, while agribusinesses may focus on actionable indicators like NDVI.

To address this diversity, dynamic data packages can be created. These are metadata-driven bundles tailored to specific user needs. For example, a researcher, policymaker, or service provider could receive the same base dataset, but with different structures, attributes, and contextual layers.

In this study, the intended audiences include developers of agricultural data platforms, researchers working on data standardization, and policy and domain experts planning the integration of drone-based monitoring systems. The ability to commercialize drone data through interoperable data spaces depends on improving metadata alignment, validating semantics, and building trusted delivery frameworks.

4 Results

This study examined how drone data processed using ODM can be transformed into a format compatible with agricultural data spaces. The findings focus on metadata standardization, semantic alignment, and integration readiness.

The comparison of metadata models showed significant overlap between DCAT, INSPIRE, and Dublin Core. This indicates that drone data products can be mapped onto standard metadata vocabularies with relatively limited effort, especially when targeting data catalogs that adopt DCAT as their base structure.

Using ODM, we produced georeferenced multispectral datasets, including RGB and NDVI imagery. These were stored in standardized formats such as GeoTIFF, accompanied by auxiliary files (e.g., JSON metadata outputs).

We developed a prototype query that validates extracted metadata terms against the AGROVOC vocabulary. The tool identified 357 terms from ODM outputs, of which 20 matched AGROVOC concepts directly. This suggests limited native compatibility with agricultural vocabularies, but also highlights the potential for semantic enrichment through mapping.

The GeoNetwork platform was tested as a metadata catalog, providing insights into structuring metadata in line with DCAT and INSPIRE. These early experiments helped identify the types of descriptors needed to improve discoverability and support FAIR principles.

Together, these results demonstrate that it is technically feasible to transform and prepare drone data for use in agricultural data ecosystems, though additional metadata processing and vocabulary alignment are needed to ensure full compatibility and reusability.

5 Discussion

The results show that drone data processed using open-source tools can be structured and enriched to meet key requirements of agricultural data spaces. However, there are important limitations that must be addressed before such data can be fully reused or commercialized.

One central challenge is metadata completeness. While ODM produces technically useful metadata files, these do not fully align with domain-specific standards. This creates gaps in interoperability that must be resolved through post-processing, enrichment, and validation.

The limited overlap with AGROVOC confirms that standard drone processing pipelines are not yet aligned with agricultural vocabularies. This restricts integration with systems that rely on semantic interoperability. The prototype query developed in this study demonstrated that vocabulary mapping is possible, but further development is needed to support dynamic, use-case-specific data packaging.

Standardized metadata alone is not sufficient. As discussed in prior sections, trusted API interfaces and clear data delivery mechanisms are also essential.

While Swagger and OpenAPI provide a robust foundation, integration into real data marketplaces requires connectors, access control layers, and legal frameworks that ensure compliance and transparency.

These findings highlight the importance of designing workflows that can adapt to the needs of different stakeholders. Farmers, researchers, and technology providers may all use drone data differently. Therefore, flexible and modular metadata structures are needed. Rather than aiming for a one-size-fits-all model, dynamic data packages that respond to user context offer a more scalable solution. Research on transforming both metadata and actual data into data products is continuing in the OurData project. The comparison of metadata fields across different standards, as shown in Table 2, requires more metadata elements and examples of different types of data products. The different development versions of this metadata schema are available at Github [55].

In future work, more attention should be given to linking drone data with auxiliary sources such as weather or soil data, which would improve value and applicability. Further testing in live agricultural data spaces is also needed to validate the technical and economic feasibility of these concepts, particularly the role of smart contracts in automating transactions and securing rights.

6 Conclusion

This study demonstrated that multispectral drone data can be processed with open-source tools and aligned with recognized metadata standards to support integration into agricultural data spaces. ODM provided a technically sound basis for generating standardized outputs, though additional metadata enrichment is required for full interoperability.

Structural similarities between standards such as DCAT, INSPIRE, and Dublin Core indicate that mapping is feasible. Prototype validation with the AGROVOC confirmed the potential for semantic alignment, despite limited out-of-the-box compatibility. Cataloging tests using GeoNetwork and API considerations based on OpenAPI/Swagger offered practical insights into publishing drone data in interoperable environments.

While promising, further work is needed to improve metadata coverage, automate semantic validation, and adapt data packages for different users. Linking drone data with contextual sources such as weather and soil information and testing smart contract mechanisms are essential next steps toward commercial scale implementation.

Acknowledgment. The authors wish to acknowledge CSC—IT Center for Science, Finland, for computational resources.

The authors thank the Ministry of Agriculture and Forestry of Finland for funding the Living Lab Data project, which has enabled research on the development of drones ecosystems toward integration with data spaces.

Special acknowledgments also to the Business Finland Our Data project, where we work together with companies to develop the data space for agriculture, frameworks [56] and data products.

References

1. European Commission: Main Initiatives in the Strategic Dialogue for the Future of EU Agriculture (2024). https://agriculture.ec.europa.eu/common-agricultural-policy/cap-overview/main-initiatives-strategic-dialogue-future-eu-agriculture_en. Accessed: 18 November 2024.
2. Lipping, T., Linna, P., & Narra, N. (Eds.) (2020). *New developments and environmental applications of drones.* Springer. https://doi.org/10.1007/978-3-030-77860-6
3. Westerlund, T., & Peña Queralta, J. (Eds.) (2023). *New developments and environmental applications of drones.* Springer (2023). https://doi.org/10.1007/978-3-031-44607-8
4. Quamar, M. M., Al-Ramadan, B., Khan, K., Shafiullah, M., & El Ferik, S. (2023). Advancements and applications of drone-integrated geographic information system technology—a review. *Remote Sensing, 15*(20), 5039. https://www.mdpi.com/2072-4292/15/20/5039; https://doi.org/10.3390/RS15205039
5. Tsouros, D. C., Bibi, S., & Sarigiannidis, P. G. (2019). A review on UAV-based applications for precision agriculture. *Information, 10*(11), 349. https://www.mdpi.com/2078-2489/10/11/349; https://doi.org/10.3390/info10110349
6. Kumar, S. P., Subeesh, A., Jyoti, B., & Mehta, C. R. (2023). *Applications of drones in smart agriculture* (pp. 33–48). https://doi.org/10.1007/978-981-19-8738-0_3
7. Eouardi, A., Latif, R., & Saddik, A. (2022). Metrics in precision agriculture using multispectral images: review and evaluation. In: K. Xiaohui, S. Elhoseny, M. Yuan (Eds.), *Distributed sensing and intelligent systems* (pp. 147–157). Springer. https://doi.org/10.1007/978-3-030-64258-7
8. Karpiński, P. (2024). The use of drones in agriculture: perspectives and limitations. In: U. Bruno, J. Lorencowicz, E. Huyghebaert (Eds.), *Farm machinery and processes management in sustainable agriculture* (pp. 219–228). Springer. https://doi.org/10.1007/978-3-031-70955-5
9. FAO (2021). Farm data management, sharing and services for agriculture development. https://doi.org/10.4060/cb2840en
10. International Organization for Standardization: ISO 19115-1:2014 (2014). Geographic Information—Metadata—Part 1: Fundamentals. https://www.iso.org/standard/53798.html. Accessed: 18 November 2024.
11. European Parliament and Council (2007). Directive 2007/2/EC Establishing an Infrastructure for Spatial Information in the European Community (INSPIRE). https://eur-lex.europa.eu/legal-content/EN/ALL/?uri=CELEX%3A32007L0002. Accessed: 18 November 2024.
12. Open Geospatial Consortium (2014). OGC SensorML: Model and XML Encoding Standard. https://www.ogc.org/standards/sensorml. Accessed: 18 November 2024.
13. International Organization for Standardization (ISO) (2024). ISO/TC 20/SC 16—Unmanned Aircraft Systems. https://www.iso.org/committee/5336224.html. Accessed: 18 November 2024.
14. Wilkinson, M.D., Dumontier, M., Aalbersberg, I.J., Appleton, G., Axton, M., Baak, A., Blomberg, N., Boiten, J.-W., Bonino da Silva Santos, L., Bourne, P.E., Bouwman, J., Brookes, A.J., Clark, T., Crosas, M., Dillo, I., Dumon, O., Edmunds, S., Evelo, C.T., Finkers, R., Gonzalez-Beltran, A., Gray, A.J.G., Groth, P., Goble, C., Grethe, J.S., Heringa, J., 't Hoen, P.A.C., Hooft, R., Kuhn, T., Kok, R., Kok,

J., Lusher, S.J., Martone, M.E., Mons, A., Packer, A.L., Persson, B., Rocca-Serra, P., Roos, M., van Schaik, R., Sansone, S.-A., Schultes, E., Sengstag, T., Slater, T., Strawn, G., Swertz, M.A., Thompson, M., van der Lei, J., van Mulligen, E., Velterop, J., Waagmeester, A., Wittenburg, P., Wolstencroft, K., Zhao, J., & Mons, B. (2016). The FAIR guiding principles for scientific data management and stewardship. *Scientific Data, 3*(1), 160018 https://doi.org/10.1038/sdata.2016.18
15. OpenDroneMap (2024). OpenDroneMap. https://www.opendronemap.org/. Accessed: 20 January 2024.
16. World Wide Web Consortium (W3C) (2020). Data Catalog Vocabulary (DCAT)—Version 3. https://www.w3.org/TR/vocab-dcat-3/. Accessed: 18 November 2024.
17. Data Space Support Centre (DSSC) (2024). Data Space Blueprint v1.5—Data Product. https://dssc.eu/space/bv15e/766065053/Data+Product. Accessed: 18 November 2024.
18. Data Space Support Centre (DSSC) (2024). Data Space Blueprint v1.5—Data, Services, and Offerings Descriptions. https://dssc.eu/space/bv15e/766069419/Data,+Services,+and+Offerings+Descriptions. Accessed: 18 November 2024
19. International Data Spaces Association (IDSA) (2024). Data Space Protocol - International Data Spaces Knowledge Base. https://docs.internationaldataspaces.org/ids-knowledgebase/dataspace-protocol. Accessed: 18 November 2024.
20. Dublin Core Metadata Initiative (DCMI) (2024). Dublin Core Metadata Element Set. https://www.dublincore.org/. Accessed: 18 November 2024.
21. Food and Agriculture Organization of the United Nations (FAO) (2024). AGROVOC Multilingual Agricultural Vocabulary. http://aims.fao.org/standards/agrovoc/functionalities/search. Accessed: 18 November 2024.
22. International Data Spaces Association (IDSA) (2024). International Data Spaces Association (IDSA). https://www.internationaldataspaces.org/. Accessed: 18 November 2024.
23. Gaia-X. https://gaia-x.eu/. Accessed: 20 January 2024.
24. Gaia-X Initiative: Agri-Gaia. (2024). https://www.agri-gaia.de//. Accessed: 18 November 2024.
25. Wamhof, T., Bernardi, A., Martini, D., Leinberger, M., Sinha, A., Tapken, H., Schliebitz, A., & Graf, H. (2023). Metadata management and asset exchange in the agricultural data ecosystem of the project agri-gaia. *Datenbank-Spektrum, 23*(2), 107–115. https://doi.org/10.1007/s13222-023-00444-3
26. Big Data Value Association: Big Data Value Association (BDVA). https://www.bdva.eu/. Accessed: 23 November 2024.
27. AgriDataSpace Consortium: D3.3: Report on data, models and interoperability solutions as technology enablers for the implementation of agricultural data space. https://agridataspace-csa.eu/deliverables/. Accessed: 23 November 2024.
28. Poikola, A., Wong, D., Lähteenoja, V., & Turpeinen, M. (2024). State of finish data spaces: experiences of pioneers in building the data economy. Sitra. https://www.sitra.fi/en/publications/state-of-finnish-data-spaces/. Accessed: 23 November 2024.
29. Noardo, F., Atkinson, R., Bastin, L., Maso, J., Simonis, I., Villar, A., Voidrot, M.-F., & Zaborowski, P. (2024). Standards for data space building blocks. *Remote Sensing, 16*(20), 3824. Multidisciplinary Digital Publishing Institute. https://www.mdpi.com/2072-4292/16/20/3824; https://doi.org/10.3390/RS16203824
30. FIWARE Foundation (2024). FIWARE: Open Source Platform for Our Smart Digital Future. https://www.fiware.org/. Accessed: 23 November 2024.

31. Otto, B., ten Hompel, M., & Wrobel, S. (Eds.) (2022). *Designing data spaces—the ecosystem approach to competitive advantage.* Springer. https://link.springer.com/book/10.1007/978-3-030-93975-5; https://doi.org/10.1007/978-3-030-93975-5
32. Simpl Programme: Simpl Initiative: Enhancing Data Interoperability Across European Data Spaces. https://simpl-programme.ec.europa.eu/. Accessed: 23 November 2024.
33. Gaia-X. International Data Spaces Association (IDSA), Big Data Value Association (BDVA), FIWARE Foundation: Data Spaces Business Alliance (DSBA). https://data-spaces-business-alliance.eu/. Accessed: 23 November 2024.
34. Food and Agriculture Organization of the United Nations (FAO) (2024). Agricultural Information Management Standards (AIMS). https://aims.fao.org/. Accessed: 23 November 2024.
35. Wolfert, S., Ge, L., Verdouw, C., & Bogaardt, M.-J. (2017). Big data in smart farming—a review. *Agricultural Systems, 153*, 69–80. https://doi.org/10.1016/J.AGSY.2017.01.023
36. European Parliament and Council (2023, December). Regulation (EU) 2023/2854 of the European Parliament and of the Council of 13 December 2023 on harmonised rules on fair access to and use of data and amending Regulation (EU) 2017/2394 and Directive (EU) 2020/1828 (Data Act). Legislative Body: CONSIL, EP. http://data.europa.eu/eli/reg/2023/2854/oj/eng. Accessed: 23 January 2024.
37. Data Spaces Radar (2024). Data Spaces Radar. https://www.dataspaces-radar.org/. Accessed: 23 November 2024.
38. Rumble Tools: Rumble Tools Drone Docking Stations. https://www.rumbletools.com/. Accessed: 26 November 2024.
39. Linna, P., Halla, A., & Suomalainen, J. (2024, May) Concept for sharing drone data in agricultural data ecosystem. In *2024 47th MIPRO ICT and electronics convention (MIPRO)* (pp. 1900–1904). IEEE. https://doi.org/10.1109/MIPRO60963.2024.10569579
40. DJI: The World Leader in Camera Drones/Quadcopters for Aerial Photography. https://www.dji.com/. Accessed: 23 November 2024.
41. Pix4D: Professional Photogrammetry and Drone Mapping Software. https://www.pix4d.com/. Accessed: 23 November 2024.
42. Optelos: Drone Data Management and AI Analytics Platform. https://optelos.com/. Accessed: 23 November 2024.
43. OpenAerialMap Project: OpenAerialMap: The Open Source Repository for Aerial Imagery. https://www.openaerialmap.org/. Accessed: 23 November 2024.
44. OpenDroneMap: WebODM: OpenDroneMap's User-Friendly Web Interface for Drone Mapping. https://www.opendronemap.org/webodm/. Accessed: 23 November 2024.
45. Barbieri, L., Wyngaard, J., Swanz, S., & Thomer, A. K. (2023). Making drone data FAIR through a community-developed information framework. *Data Science Journal, 22*(1). https://doi.org/10.5334/DSJ-2023-001
46. European Commission: DCAT-AP Validation Service. https://www.itb.ec.europa.eu/shacl/dcat-ap/upload. Accessed: 26 November 2024.
47. JSON Schema: A Vocabulary for JSON Validation. https://json-schema.org/. Accessed: 23 November 2024.
48. W3C: Simple Knowledge Organization System (SKOS). https://www.w3.org/TR/skos-reference/. Accessed: 23 November 2024.
49. VocBench Development Team: VocBench: Collaborative Environment for Managing Controlled Vocabularies. http://vocbench.uniroma2.it/. Accessed: 23 November 2024.

50. Apache Software Foundation: Apache Jena: Semantic Web Framework. https://jena.apache.org/. Accessed: 23 November 2024.
51. PyCSW Project: PyCSW: OGC Compliant Metadata Catalogue Service. https://pycsw.org/. Accessed: 23 November 2024.
52. GeoNetwork Opensource Project: GeoNetwork: Opensource Geospatial Metadata Catalogue. https://geonetwork-opensource.org/. Accessed: 23 November 2024.
53. GraphQL (2015). A Query Language for APIs. Meta Platforms, Inc. (formerly Facebook). https://graphql.org/. Accessed: 26 November 2024.
54. Swagger (2016). OpenAPI Specification Framework. OpenAPI Initiative. https://swagger.io/. Accessed: 26 November 2024.
55. Petri Linna: metadata (GitHub repository) (2025). https://github.com/petrilinna/metadata. Accessed: 20 December 2024.
56. Pussi, K., Linna, P., Suomi, P., Kaustell, K., & Pesonen, L. (2025). Reference framework for metadata description to commensurate data in grain production. *Data in Brief, 60*, 111557. https://doi.org/10.1016/j.dib.2025.111557

Drone Logistics and Infrastructure Integration

Conceptual Development of an Autonomous All-Electric Wing-in-Ground Effect Cargo Craft

Jussi Aaltonen[1](✉), Akseli Arola[1], Peetu Tuominen[1], Kalle Hakonen-Milošević[1], Zorana Hakonen-Milošević[2], and Kari Koskinen[1]

[1] Mechatronics Research Group, Tampere University, Tampere, Finland
jussi.aaltonen@tuni.fi
[2] CAR Universidad Politécnica de Madrid—CSIC, Madrid, Spain

Abstract. The concept of ground effect, where a craft flying close to a surface experiences an improved lift-to-drag ratio, has been known since the early twentieth century. Finnish engineer Toivo Kaario pioneered this field in the 1930s. The Soviet Union advanced wing-in-ground (WIG) technology with ekranoplans like the Caspian Sea Monster and Lun-class. In the West, Alexander Lippisch's reverse delta wing design allowed for stable ground effect flight. Modern developments, such as the REGENT Seaglider, have further advanced the field. Recent innovations, including electric propulsion, have enhanced WIG performance and efficiency while reducing environmental impact. Despite these advancements, regulatory and operational challenges remain. The Horizon Europe project AIRSHIP aims to develop and demonstrate an autonomous all-electric WIG craft for cargo transport in archipelagoes, inland seas, and inland waterways. The project seeks to address fundamental challenges of WIG crafts through optimized aerodynamics and advanced control systems, fulfilling market demands for efficient and versatile transport solutions. The project will produce three prototypes, each more advanced than the previous, culminating in a final technology demonstrator prototype capable of autonomous missions in real operational environments. This paper encapsulates the research findings and outcomes of the conceptual development phase.

Keywords: Wing in ground effect (WIG) · Autonomous craft · Cargo transport · Aerodynamic analysis · Stability analysis · Take-off performance · Power estimations · Floatwing design · Ground effect · Operational requirements

1 Introduction

The ground effect phenomenon has been recognized since the early twentieth century. Significant advancements in studying and developing vehicles based on this phenomenon were made by two pioneers during the pre- and post-war years.

The earliest pioneer in experimenting with ground effect vehicles was Finnish engineer Toivo Kaario. Kaario built his first prototype in 1935 and applied for a patent on January 2, 1935 [13]. Thus, he is regarded as the inventor of the ground effect vehicle. The first scientific studies on ground effect were conducted by German-born engineer

M. Laakkonen (ed.), *DroneAI for Resilient Nature*,
https://doi.org/10.1007/978-3-032-19008-6_12

Alexander Lippisch, beginning in the early 1920s. However, Lippisch began experimenting with ground effect vehicles only in the post-war 1950s. He patented his so-called reverse delta wing in 1964 [15]. The reverse delta wing has since become the most widely adopted wing configuration in ground effect vehicles.

The most notable development in the field of ground effect vehicles occurred in the Soviet Union from the 1960s to the 1990s. Various development programs led to a variety of full-scale prototypes, which were extensively tested and demonstrated in the Caspian Sea [21]. Notably, the Soviet development pathway was based on Kaario's work rather than Lippisch's.

Flying in ground effect reduces drag and increases lift compared to flying out of it. Drag is decreased due to the weakening of airflow vorticity around and in the slipstream of the wing. Lift is increased due to restricted downwash, i.e., air compression under the wing. These phenomena occur in proximity to a physical surface, typically no more than one wingspan above it. This unique aerodynamic condition allows ground effect vehicles to potentially achieve higher fuel efficiency and extended range than conventional aircraft flying at higher altitudes [5, 10].

However, ground effect also presents design challenges, such as the need for precise control over altitude and stability, especially during take-off, landing, or in varying weather conditions during cruise. Additionally, vehicle geometry, including wing shape and the flight altitude-to-wingspan ratio, plays a critical role in optimizing these aerodynamic benefits. Consequently, ground effect vehicles are particularly suitable for missions over large bodies of water, where the surface remains relatively smooth, allowing for more consistent performance [20].

2 AIRSHIP Project

The AIRSHIP project represents a significant advancement in the field of autonomous cargo transport, leveraging the unique capabilities of Wing in Ground Effect (WIG) technology. WIG crafts operate by flying close to the surface, typically over water, where they benefit from increased lift and reduced drag due to the ground effect. This phenomenon allows WIG crafts to achieve higher transport efficiency than aircraft, making them an attractive solution for short to medium-range cargo transport [4, 5, 21].

The primary objective of the AIRSHIP project is to develop and demonstrate an autonomous WIG craft specifically designed for cargo transport across sea and inland waterways.

The development process of the AIRSHIP project is structured into several phases, each focusing on different aspects of the WIG craft's design, analysis, and testing. The project will produce three prototypes, each more advanced and complex than the previous one, culminating in a final prototype capable of performing autonomous missions in real operational environments. Digital twins will be employed to simulate the prototype operations and validate performance before test flight (Fig. 1).

The conceptual design phase, detailed in this article, focuses on establishing the operational requirements, conducting engineering analyses, and defining the aerodynamic and structural configuration of the WIG craft. The design approach integrates state-of-the-art analytical methods, numerical simulations, and best practices from aircraft and

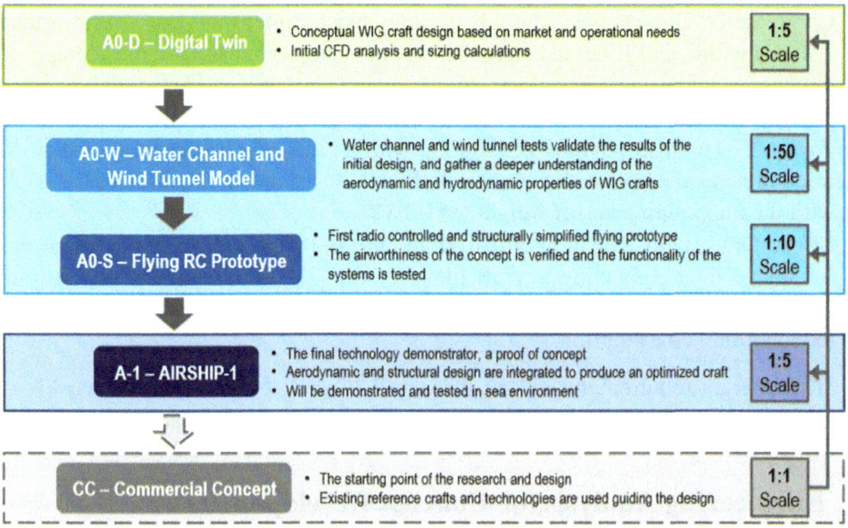

Fig. 1 Airship prototypes and their purposes in the development process

marine vessel design. The goal is to create a WIG craft that meets the operational needs of cargo transport while maximizing efficiency, stability, and maneuverability.

3 Technical Requirements

The technical requirements are critical in defining the capabilities and performance criteria for the autonomous Wing in Ground Effect (WIG) craft. Each category addresses different aspects of the craft's performance and operational environment, ensuring that the WIG craft can meet both basic functionality and competitive performance standards.

3.1 Target Requirements

The Operational Target Requirements are designed to make the WIG craft competitive with existing transportation methods. These requirements focus on enhancing the craft's performance and operational efficiency to meet market demands. Key parameters include:

- Wave Height Tolerance: The craft must be able to tolerate a maximum wave height of 3.30 m during cruise. This tolerance is based on Europe's multi-decade maximum mean significant wave height (SWH) of 2.20 m, expected in the sea surrounding the Canary Islands during winter. An efficient cruise SWH value of 1.30 m represents the calmer conditions of the Mediterranean and Baltic Seas [1, 6].
- Effective Range: The target effective range is 1000 km when fully loaded. This range reflects the longest commercially viable flight routes in the European archipelagos, such as from continental Europe to the Macaronesia archipelago [16].

- Cargo Space Dimensions: The target cargo space dimensions are 8.0 m in length, 2.2 m in width, and 1.9 m in height. These dimensions are based on the average cabin dimensions of business jets like the Gulfstream G280 [12] and Dassault Falcon 2000 [7], ensuring accommodation for high-value cargo or passengers.
- Payload: The target payload is 4300 kg, with a payload mass fraction of 0.40. This fraction measures the craft's structural efficiency, balancing the mass of the payload with the maximum take-off weight (MTOW).
- Lift-to-Drag Ratio: The target lift-to-drag ratio is 18.5, which provides a reasonable advantage over competing aircraft like the ATR 72-600F, which has a lift-to-drag ratio of 16.7 [3]. A higher lift-to-drag ratio indicates better aerodynamic efficiency, reducing fuel consumption.

These target requirements ensure that the WIG craft can compete effectively with existing transportation methods, offering superior range, cargo capacity, and efficiency.

4 Engineering Analysis for Concept Development

The engineering analysis for the conceptual development involves a comprehensive evaluation of various design aspects to ensure the WIG craft meets the operational requirements and performance targets.

4.1 Aerodynamic Analysis

The lift-to-drag ratio (*L*/*D*) indicates the overall aerodynamic efficiency of the craft, and thus needs to be maximized. A graphic representation of this process, i.e., how the aerodynamic efficiency affects the design problem can be seen in Fig. 2.

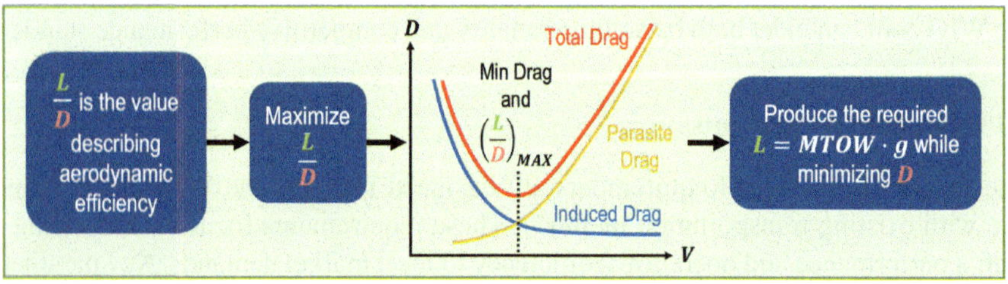

Fig. 2 Aerodynamic efficiency driven design process

The maximization of the *L*/*D* can be done either by maximizing lift or minimizing drag. The approach for maximizing lift was found to be inferior to minimizing drag. As lift of the craft must be equal to its weight to maintain steady level cruise flight, increase in lift enables carrying heavier payload. However, in commercial applications the absolute carrying capacity of the vehicle is irrelevant compared to the cost of carrying the payload. In consequence, the maximization of *L*/*D* is more desirable by focusing on the minimization of the drag than maximizing the payload capacity.

To demonstrate the drag curves for the CC craft the total drag equation is derived. The drag of the craft D_{tot} can be written as [14]:

$$D_{\text{tot}} = \frac{1}{2}\rho {v_c}^2 C_D S \tag{1}$$

where ρ is the air density at ISA conditions, v_c is cruise speed and S is wing area. Drag coefficient C_D is defined as the drag polar [14]:

$$C_D = C_{D_0} + C_{D_i} \tag{2}$$

where C_{D_0} is zero lift drag coefficient. At this point of the project the best estimate for $C_{D_0} = 0.0306$, which is a good approximation for a small regional transport aircraft [8]. For a wing in free flow, the induced drag coefficient C_{D_i} is defined by aspect ratio AR and lift coefficient $C_{L\text{OGE}}$ as [14]:

$$C_{D_i} = \frac{{C_{L\text{OGE}}}^2}{\pi e \text{AR}} \tag{3}$$

where e is Oswald's efficiency factor. Lift coefficient is defined as [14]:

$$C_{L\text{OGE}} = \frac{\text{mg}}{\frac{1}{2}\rho {v_c}^2 S} \tag{4}$$

where m is the total mass, g is gravitional acceleration, ρ is thc air density at ISA conditions, v_c is cruise speed and S is wing area.

The total drag curves for a 14000 kg craft with various wing aspect ratio AR and wing area S combinations can be established. The curves use constant values of $\rho = 1.225$ kg/m^3 and $g = 9.807$ m/s^2 together with Oswald's efficiency factor $e = 0.9$ as inputs [18]. The curves are presented in Fig. 3.

It can be seen from Fig. 3 that increasing AR significantly decreases total drag of the wing, while increasing wing size doesn't significantly affect drag. Increase of AR also results in the total drag curve to move left and down, i.e. the least total drag is generated at lower speed.

Maximizing L/D requires the examination of drag polar curves in ground effect (IGE) and determining the required lift for feasible operational capability. The induced drag IGE $C_{D_{i,\text{IGE}}}$ can be estimated by Phillips [17]:

$$C_{D_{i,\text{IGE}}} = \frac{{C_L}^2}{\pi e \text{AR}} \mu_D \tag{5}$$

where C_L is lift coefficient, e is oswalds coefficient, AR is aspect ratio and μ_D is a parameter that describes ground effect (GE) influence according to spanwise relative flight altitude h/b. For untwisted tapered wings at small angles of attack, the parameter model is defined from Prandtl's lifting-line theory and evaluated against numerical solutions as:

$$\begin{aligned}\mu_D = 1 - \left(1 - 0.157\left(\text{TR}^{0.757} - 0.373\right)\left(\text{AR}^{0.417} - 1.27\right)\right)e^{\left(-4.74\left(\frac{h}{b}\right)^{0.814}\right)} \\ - \left(\frac{h}{b}\right)^2 e^{\left(-3.88\left(\frac{h}{b}\right)^{0.758}\right)}\end{aligned} \tag{6}$$

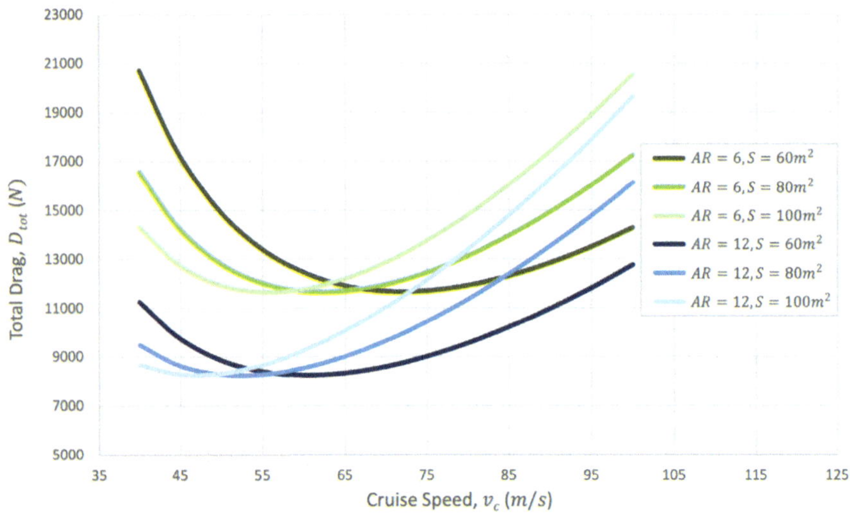

Fig. 3 The effect of different AR and *S* combinations for the drag curve

where TR is taper ratio, AR is aspect ratio and e is the Napier's constant. As can be seen from comparing Eqs. 4 and 5, increasing wingspan and thus AR reduces induced drag both IGE and OGE. However, increasing span IGE also reduces spanwise relative flight altitude *h*/*b* which leads to further reduction of the induced drag. Thus, it was concluded that one of the design objectives should be maximizing AR. Relation AR/μ_D can be thought as an effective AR of a wing flying IGE, i.e. the wing IGE behaves as if its AR was greater than it geometrically is. This has similar effect on induced drag as actual increase of AR would have, which is demonstrated in Fig. 4.

While induced drag is decreased IGE, lift generated is increased by equation [17].

$$C_{L_{\mathrm{IGE}}} = C_{L_{\mathrm{OGE}}} \mu_L \tag{7}$$

where the OGE to IGE transformation parameter μ_L for SGE is written as:

$$\mu_L = 1 + (1 - 2.25\left(\mathrm{TR}^{0.00273} - 0.997\right)) \left(\mathrm{AR}^{0.717} + 13.6\right) \frac{288\left(\frac{h}{b}\right)^{0.787} e^{\left(-9.14\left(\frac{h}{b}\right)^{0.327}\right)}}{\mathrm{AR}^{0.882}} \tag{8}$$

where the parameters are same than the ones used in the Eq. 6. Using the equations presented, the effect of GE on the wing's drag is demonstrated in Fig. 4. The drawn curves use constant values $\rho = 1.225$ kg/m^3 and $g = 9.807$ m/s^2 together with the Napier's constant (Euler's number) $e = 2.718$ as inputs. The craft related parameters used are $m = 14\,000$ kg, Oswald's efficiency factor $e = 0.9$ and $h/b = 0.1$.

From Fig. 4, it is seen that flying IGE has similar effect than to increase in AR. Figure 4 shows that the effective AR doubles as the wing area is kept constant. In practice, this gives two design options: To reduce the AR and still gain efficiency of

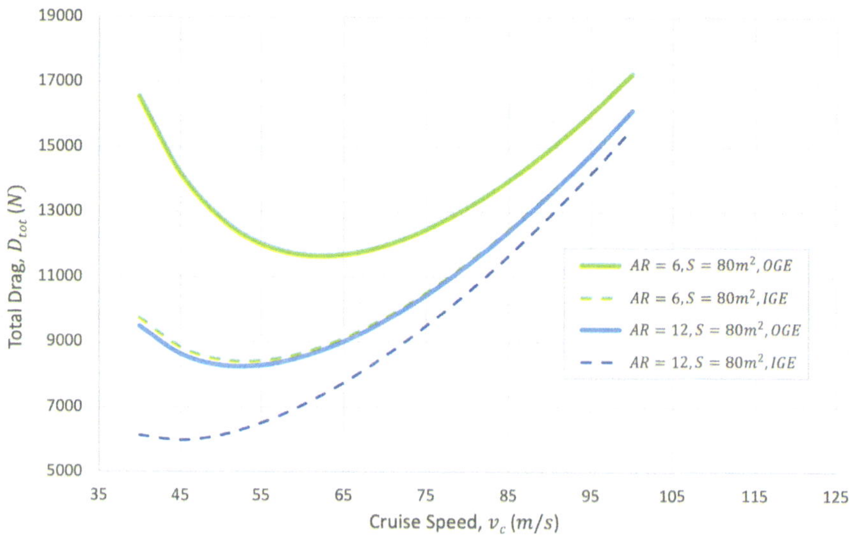

Fig. 4 The effect of GE on the total drag curve

a significantly higher AR wing, or to keep the AR as high as possible to reduce the drag even further. While having as high AR as possible would be best from the purely aerodynamics standpoint, the structural integrity and stiffness required set the upper limit for AR.

Computational Fluid Dynamics (spa) simulations were conducted using ANSYS Fluent 2022 software to evaluate different feasible span-dominated- (SGE) and chord-dominated-ground effect (CGE) wing configurations [2]. The analysis considered five wing configurations: Reverse delta wing (CGE), compound wing (CGE), high wing (SGE), mid wing (SGE), and floatwing (SGE) (Fig. 5). All wings have same profile (NACA 4421) and all wings was dimensioned so that craft can bank 10° without increasing altitude and that either $h/b \leq 0.1$ (SGE) or $h/c \leq 0.1$ (CGE). The model was constructed so that absolute flying altitude (measured as clearance between the surface and the lowest point of the wing), and velocity are identical [9].

Each configuration was assessed for its aerodynamic performance, including lift, drag, and overall efficiency. The CFD analysis revealed that the floatwing configuration provided the best aerodynamic performance. It achieved the highest L/D ratio, indicating superior efficiency. The floatwing design integrates the floating and lifting functions, reducing drag and enhancing lift generation. The key parameter values for the mentioned configurations are summarized in Table 1.

Spalart–Allmaras turbulence model for viscous flow was used. The reason for using the Spalart–Allmaras turbulence model is its simplicity, computational efficiency, and suitability for external aerodynamics, which makes it often preferred over models such as L-VEL, *yplus*, k–ε, k–ω, and V2F. Relatively coarse mesh of 5–10 million cells was used, but to ensure a reasonable accuracy for the analyses, the convergence criterion was set to Δnut < 10^{-6} [2].

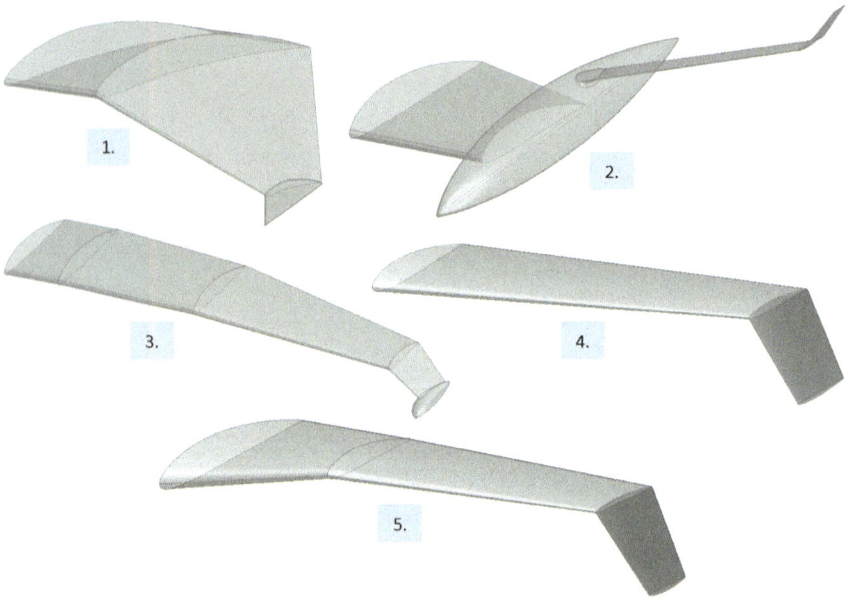

Fig. 5 Wing configurations analyzed with CFD. 1—Reverse delta wing (CGE), 2—Compound wing (CGE), 3—High wing (SGE), 4—Mid wing (SGE) and 5—Floatwing (SGE)

Table 1 Key parameter values of the performed CFD analysis

	Wing configuration	Clearance (m)	Wing area (m^2)	Velocity (m/s)	*L*/*D*	*L* (N)
1.	Reverse delta wing	0.92	102	65	17.1	117000
2.	Compound wing	0.92	104	65	24.0	184000
3.	High wing	0.92	79	65	34.9	150000
4.	Mid wing	0.92	76	65	39.2	132000
5.	Floatwing	0.92	76	65	40.5	134000

All wings were analyzed with the aerodynamic effects of endplates are included. Endplates are typical WIG wing configuration features mounted on the wing tip to decrease induced drag by blocking airflow leakage from under the wing [20]. Compound wing was analyzed including its float due to it functioning as an endplate similar way as in the other wing configurations [22]. Similarly, floatwing has the floating configuration integrated into the root section of the wing by default. Therefore, due to aerodynamically sub-optimal fuselage shape needed for the floating configuration in rest of the configurations, they will gain an additional disadvantage in regard of *L*/*D*.

The floatwing configuration's superior performance is attributed to its ability to minimize drag while maximizing lift. The integration of floating and lifting functions reduces the overall drag, making it the most efficient design. This configuration was selected for further development based on its aerodynamic advantages.

4.2 Stability Analysis

Generally, an airplane is considered longitudinally stable when two separate requirements are met [14]. The first requirement states that:

$$C_{M_{\mathrm{AC}}} > 0 \tag{9}$$

which means the moment coefficient at the aerodynamic center (AC) $C_{M_{\mathrm{AC}}}$ must be positive. The second requirement states that:

$$\frac{\mathrm{d}}{\mathrm{d}\alpha} C_{M_{\mathrm{CG}}}(\alpha) < 0 \tag{10}$$

which means that the gradient of the moment coefficient at center of gravity (CG) $\frac{\mathrm{d}}{\mathrm{d}\alpha} C_{M_{\mathrm{CG}}}(\alpha)$ must be negative. These two requirements ensure that the airplane has a tendency of returning into equilibrium state after disturbance in angle of attack α. To achieve this, CG must be in front of the AC. Using aerodynamics analysis software OpenVSP it could be shown that the craft is stable OGE [9]. The two requirements are met as shown in Fig. 6, respectively.

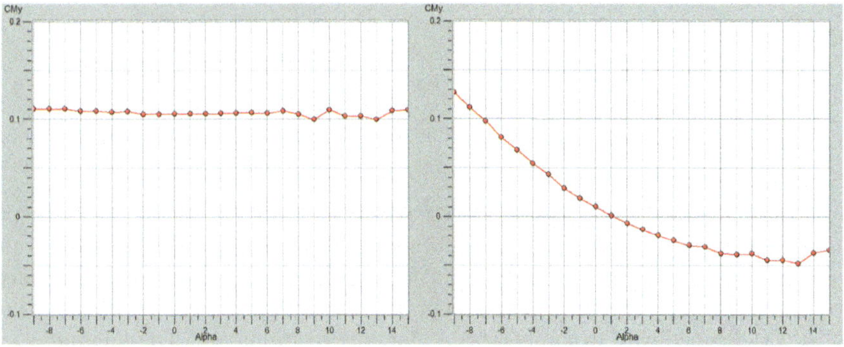

Fig. 6 C_M at the aerodynamic center (right) and C_M at the center of gravity (left)

CG of a seaplane should lay withing 10°–15° angle in front of the hull bottom step [11]. This translates to approximate locations of 0.66 m forward from the trailing edge of the root chord. The AC lays 0.40 m behind the CG, which is equivalent to static stability margin of 12.3%, which is satisfactory, although the higher the static margin, the more stable the craft would be.

The stability analysis also considered additional requirements for WIG crafts, such as Idorov's criterion and binding criterion, which will be further explored in the prototype development phase.

4.3 Take-off Performance

The wing area S is largely determined by the required take-off performance, and more precisely, wing loading, take-off speed and maximum lift coefficient. Largest wing loading that the craft can take-off with is dictated by relation:

$$\left(\frac{m}{S}\right)_{\max} = \frac{\frac{1}{2}\rho {v_{\mathrm{to}}}^2 C_{l_{\max}}}{g} \tag{11}$$

Take-off speed v_{to} with different $C_{l_{\max}}$ values is shown in Fig. 7.

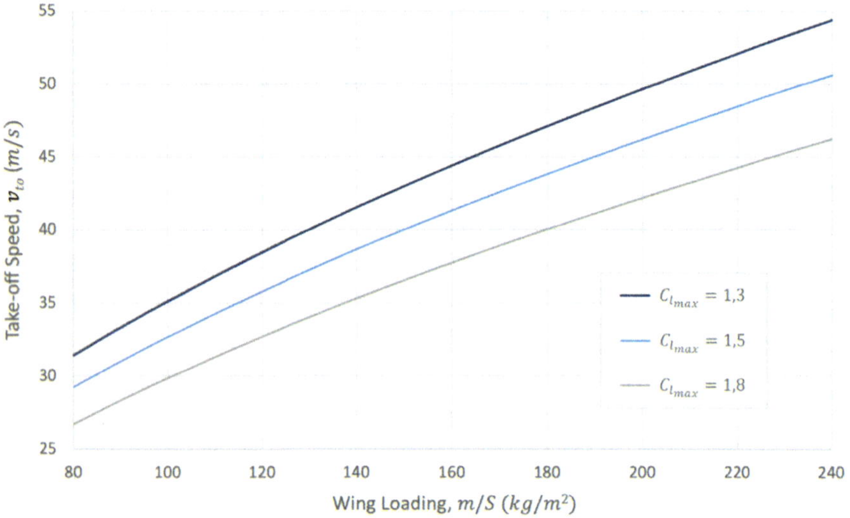

Fig. 7 Take-off speed relation to wing loading and maximum lift coefficient

To make sure the craft can take-off controlled and safe manner the take-off speed was limited according to Richards [19] and similar seaplane, e.g. Beriev Be-103, values to 140 km/h (38.9 m/s) during calm seas i.e. for sea states 0–2. Preliminary CFD analysis showed that $C_{l_{\max}} = 1.5$ is attainable, though somewhat higher values are possible. From this, it was deemed wise to limit $(\mathrm{m/S})_{\max}$ initially to 150 kg/m^2. The minimum wing area required for take-off can be approximated using relation:

$$S_{\min} = \frac{m_t}{\left(\frac{m}{S}\right)_{\max}} \tag{12}$$

The analysis showed that a wing area of 70–110 m^2 is required for the take-off. The take-off speed was limited to 140 km/h during calm seas, with a maximum lift coefficient $\left(C_{l_{\max}}\right)$ of 1.5.

The take-off performance analysis highlights the importance of optimizing wing loading and lift coefficients to ensure safe and efficient take-off. The incorporation of high lift devices and hydrofoils can significantly improve take-off capabilities.

4.4 Power Estimations

Preliminary power requirement was estimated for different stages of flight, including displacement mode and hydroplaning mode while taxiing, actual take-off, and cruise flight.

Aerodynamic drag at various stages of flight was estimated using equations presented earlier. Hydrodynamic resistance at displacement mode was assumed to behave as presented for a floatwing [19]. In hydroplaning mode, a rough estimate of the hydrodynamic resistance was got by estimating the submerged part of the wing acting as a small wing in water.

When total drag D_{tot} is known at each stage of operation, power required P can be derived using the equation:

$$P = \frac{vD_{tot}}{\mu_{prop}} \tag{13}$$

Where v is the speed variable through the stages and μ_{prop} is propeller efficiency. Combined drag generated and power required for 1:5 scale A-1 prototype is estimated in Fig. 8 [19].

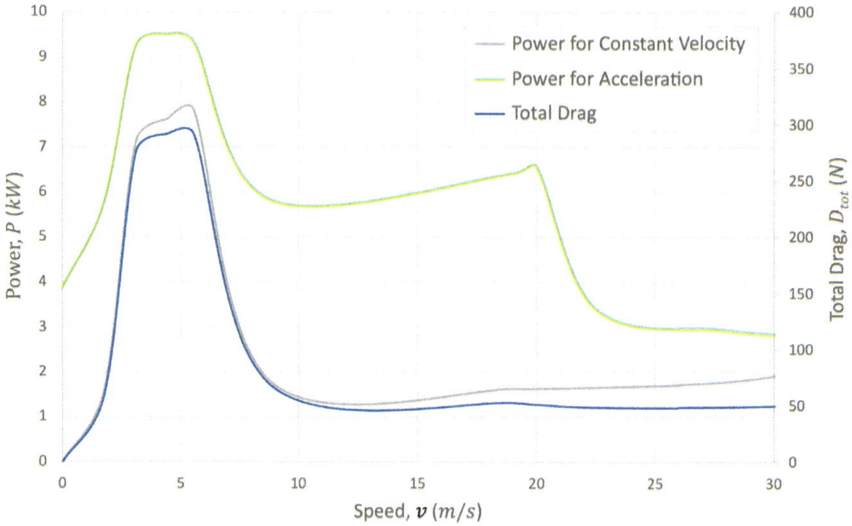

Fig. 8 Power and drag curves of A-1 at different stages of flight

The power required for take-off was found to be up to five times higher than that needed for cruise flight (Fig. 8). The peak power requirement during take-off is due to high drag and low propeller efficiency at low speeds.

The power estimations underline the power demand during take-off, necessitating efficient propulsion systems. The use of water jet propulsion units can help meet these demands, improving take-off performance and waterborne maneuvrability.

As can be seen from Fig. 8 the power required for take-off run at its peak may be as much as five times of what is required for cruise flight. This is mostly due to poor efficiency of propeller at low air speeds and high drag before hydroplaning threshold is reached at around 7 m/s. From the power for acceleration curve can be estimated that the time for take-off (from zero to $v_{to} = 19$ m/s) is 15.8 s. If necessary, the take-off can be assisted with a water jet propulsion unit, which would also greatly improve watermanoeuvrability.

The drag curve ignores the effect of water sprays, which can be a significant portion of the total drag during a high-speed take-off run. To account for unexpected sources of drag the motors should be dimensioned to produce moderately more power than Fig. 8 indicate. Actual hydrodynamical properties of the floatwing will later be tested in water channel tests.

4.5 Configuration Development Problem

The development problem description outlines the practical guidelines and requirements that need to be fulfilled to meet the design objectives. These include:

1. **Flight Altitude**: Verify that the flight altitude is sufficient for staying in significant ground effect (GE) region.
2. **Lift-to-Drag Ratio**: Maximize the L/D ratio while maintaining acceptable cruise speed.
3. **Streamlining**: Make the configuration clean by streamlining and reducing frontal area.
4. **Lift Production**: Produce enough lift for take-off while optimizing lift production for efficient cruise flight.
5. **Floating Configuration**: Provide a feasible floating configuration that produces large enough floating volume while minimizing hydrodynamic drag.
6. **Structural Integrity**: Ensure the wingspan is sufficient for necessary structural integrity and stiffness regarding GE-specific load profile.
7. **Rolling Capability**: Set wingspan in relation to flight altitude sufficient for necessary rolling capability.

These guidelines ensure that the WIG craft meets the aerodynamic, structural, and operational requirements necessary for successful deployment in real-world environments.

5 Aerodynamic and Structural Configuration

When combining the design objectives, operational limit, and target requirements with the results of the CFD analysis and development problem description, it can be concluded that a high AR floatwing is the most feasible choice for further development. Together with the floatwing choice, there comes other significantly beneficial features that can be incorporated in the configuration. Figure 9 shows the most weighting key factors and their correspondence for the choice.

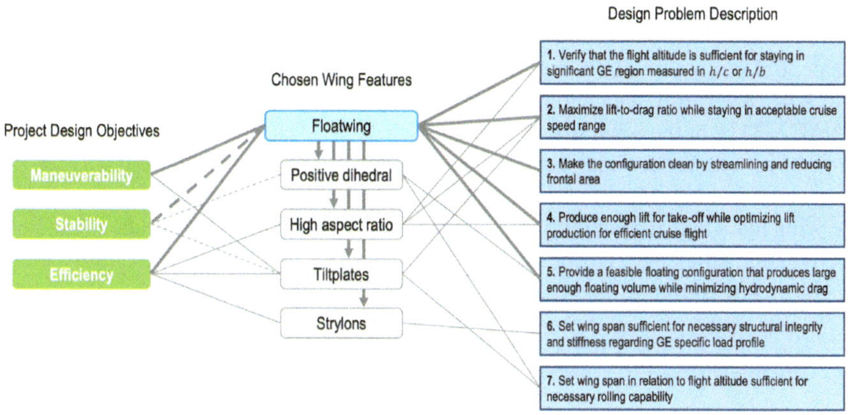

Fig. 9 Chosen wing configuration fulfilling the overall project design layout

In Fig. 9, the main influences are shown by thicker connecting lines and the effect of the integrated and additional features of the floatwing on the design layout by thinner connecting lines. Dashed lines indicate larger uncertainty due to the lack of available previous research data on the subject.

The chosen configuration has the following main components: main wing, fuselage, tail, central pylon, two outer pylons, and two motor nacelles. The wing is mounted under a conventional fuselage by three pylons. As the two outer pylons, named as "strylons", have similar structural function to wing struts. The strylons are also used to mount the motor nacelles. Wingtips have winglet-like endplates, which are named as "tiltplates". The tail is π-shaped. WIG craft's chosen configuration is illustrated in Fig. 10.

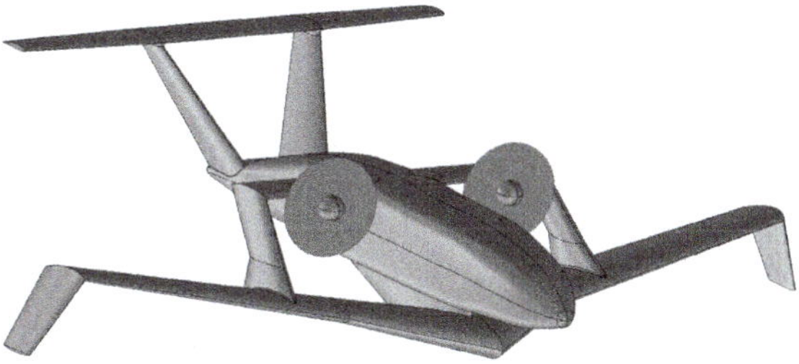

Fig. 10 The configuration of AIRSHIP WIG craft

The configuration is relatively close to the shape of a traditional aircraft, which improves the OGE flying capability if needed by collision avoidance or optimal mission planning. The targeted 1000 km range is also more probable to be reached with an high AR configuration. The *L*/*D* needs to be high enough for the required capacity of the batteries, hydrogen, or a hybrid power system.

5.1 Lifting Surfaces

The central section of the floatwing serves as the main floating body of the craft. It uses NACA 4418 aerofoil, which thins down to NACA 4416 at the central section and further to NACA 3412 at the wing tip, and finally NACA 2410 at the tip of the tiltplate. These aerofoils provide a flat undersection that performs well in ground effect (IGE). The outer wings provide additional lift, while the tiltplates improve aerodynamic performance IGE. The wing has a high aspect ratio of 7.5, which is very high for a WIG craft, enhancing its aerodynamic efficiency. The positive dihedral provides possible lateral stability, helping the craft resist rolling.

5.2 Strylons and Motor Placement

Placing the motors in strylons minimizes interference and increases clearance from the water surface, reducing the effects of water spray during take-off and landing. The strylons provide additional support to the wing, enhancing its structural integrity. The strylons can be designed to be flexible to absorb and dampen vibrations, reducing stress concentrations.

5.3 Fuselage

The fuselage is designed to minimize drag, with a slightly lowered nose and squared-off sides to eliminate downdraft at ground effect. The fuselage provides multiple possible placement positions for access hatches, such as the cargo door, ensuring flexibility in cargo handling.

The streamlined design reduces drag, improving aerodynamic performance. The optimized shape enhances the craft's overall efficiency. The multiple placement positions for access hatches provide flexibility in cargo handling.

5.4 Prototype Specifications

The main design parameters for the commercial concept (CC) craft and the prototypes are summarized in Table 2. These parameters are derived from the data presented and in accordance with the established theoretical framework, employing estimations and fundamental analyses guided by standard equations regarding aerodynamics, hydrodynamics, and aircraft performance.

The chosen aerodynamic and structural configuration of the WIG craft is designed to meet the operational requirements and performance targets, ensuring efficient, stable, and competitive performance in real-world environments. The subsequent phases of the project will involve further refinement and optimization of this configuration, leading to the development and testing of prototypes.

6 Autonomous Control

Autonomous control and the flight control system in general are generally not in the scope of conceptual design phase, but as AIRSHIP craft is specifically designed to be autonomous, they are briefly discussed here.

Table 2 Specifications of Airship prototypes A0-S and Airship-1, and commercial concept craft

Parameter	Commercial concept	AIRSHIP-1	A0-S
Linear scale factor	1.0	0.2	0.1
Wingspan (m)	25.0	5.0	2.5
MTOW (kg)	12,000–16,000	100–120	10–14
Payload mass (kg)	4000–7000	20	–
Cruise speed (km/h)	210–234	97–108	38–47
Cruise motor power (kW)	300–400	1.7–2.2	–
Take-off speed (km/h)	120–140	65–72	–
Take-off motor power (kW)	1000–1200	9.0–10.0	–

The basic configuration of the AIRSHIP craft does not impose specific requirements on the flight control system or autonomous control architecture. The design prioritizes inherent stability, and it lacks features that would significantly enhance or hinder controllability. Although electric propulsion systems enable precise and rapid thrust modulation, employing propulsion control as a primary method for directional control in fixed-wing aircraft presents several limitations:

- The effectiveness of differential thrust, or propeller braking is highly dependent on flight conditions such as airspeed, altitude, and angle of attack, making it unreliable across the full operational envelope.
- Propulsion-based control typically lacks the authority of conventional aerodynamic surfaces, especially during high-speed flight where aerodynamic forces dominate.
- Reliance on propulsion for control can compromise redundancy and safety, particularly in the event of motor failure or power system anomalies. Operating high-speed craft at a very low altitude leaves zero margin for any errors or malfunctions, which sets overall dependability of the control system priority.

The control system of the AIRSHIP craft is designed to enable safe, efficient, and autonomous operation while maintaining human supervision and oversight. The human-in-the-loop concept is important as it ensures that while the vehicle operates autonomously, a pilot or operator always retains ultimate authority and can intervene if necessary. The control architecture is outlined in Fig. 11.

The control system consists of three key components:

1. Ground Control Station (GCS): The GCS is responsible for mission planning and monitoring vehicle status. It is the interface between the human operator and the vehicle's onboard systems.
2. Autonomous Flight Management System (AFMS): The AFMS serves as the "brain" of the craft. The AFMS autonomously handles three critical functions: guidance, by determining the optimal and safe trajectory to meet mission objectives; navigation, by tracking the vehicle's position and orientation using sensor data; and mission execution, by managing high-level tasks such as reaching waypoints and maintaining

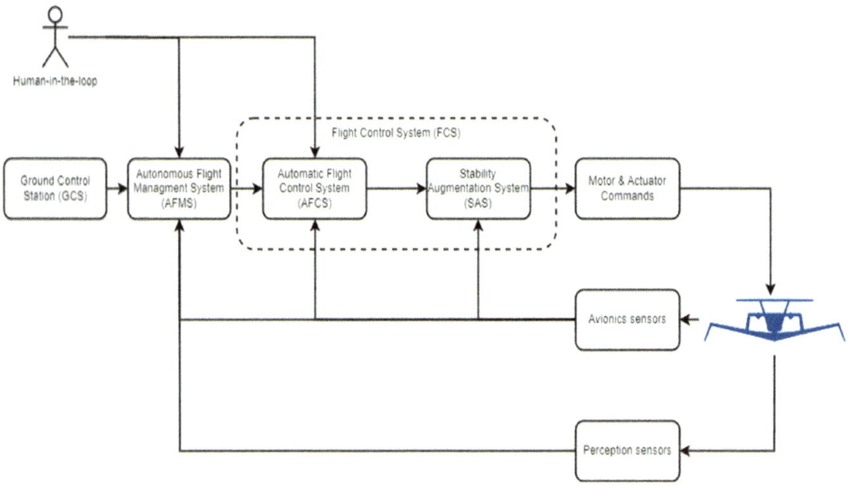

Fig. 11 Autonomous flight control system functional block diagram

operational parameters. Once a safe and stable trajectory is computed, the AFMS sends it to the FCS for execution.

3. Flight Control System (FCS): The FCS executes the trajectory provided by the AFMS by controlling the actuators and propulsion systems while ensuring that it operates within its flight envelope—the range of speeds, altitudes, and conditions in which it can safely function. It consists of two main sub-systems:

   - Automatic Flight Control System (AFCS): Responsible for translating high-level commands into precise and coordinated movements of the control surfaces and motors. The AFCS relies on predefined control laws to govern how inputs are translated into outputs for controlling pitch, roll, and yaw. These control laws include normal mode, which ensures optimal performance during standard operations, and direct mode, a fail-safe mode that provides basic manual-like control in case of system malfunctions. The AFCS continuously ensures that control actions remain within the defined flight envelope to avoid unsafe conditions.
   - Stability Augmentation System (SAS): Works in tandem with the AFCS to dampen oscillations, suppress undesirable motions (e.g., turbulence-induced pitch or roll), and ensure overall stability during flight.

The unique dynamics of IGE operation introduce specific challenges to the control system. The AFCS must maintain a precise altitude to stay within IGE whenever possible to ensure the maximum operational efficiency while continuously adapting to varying wave heights. Additionally, a smooth transition between waterborne and airborne states during take-off and landing is critical for a stable operation.

7 Conclusions

This article presented detailed the conceptual development phase of a cargo transport WIG-craft, focusing on the operational requirements, engineering analyses, and the chosen aerodynamic and structural configuration. The main findings and contributions of this phase are summarized below.

The design approach for the WIG-craft is driven by the need to maximize aerodynamic efficiency while meeting the operational requirements for cargo transport. The process involved a combination of analytical calculations, numerical CFD analysis, and best practices from aircraft and marine vessel design. This comprehensive approach ensured that the design is both innovative and practical, capable of meeting the demands of modern cargo transport.

The operational requirements were divided into minimum operational limit requirements and operational target requirements. The minimum requirements ensure that the craft can operate safely and effectively under typical conditions, while the target requirements aim to make the WIG craft competitive with existing transportation methods. Key parameters such as effective range, cargo space dimensions, wave height tolerance, payload capacity, and lift-to-drag ratio were defined to guide the design process.

The chosen configuration for the WIG-craft is the floatwing design, which integrates the floating and lifting functions into a single structure. This design offers several advantages, including reduced drag, increased lift, and improved overall efficiency. The placement of motors in strylons minimizes aerodynamic interference and enhances structural integrity. The streamlined fuselage design further reduces drag and improves aerodynamic performance. The main design parameters for the commercial concept and prototypes were established, providing a clear framework for further development.

The stability analysis confirmed that the WIG-craft is longitudinally stable out of ground effect, with a satisfactory static stability margin. This stability is crucial for maintaining a steady flight path and ensuring safe operation. Additional stability criteria specific to WIG crafts, such as Idorov's criterion and binding criterion, will be addressed in the next phases of the research.

The take-off performance analysis highlighted the importance of optimizing wing loading and lift coefficients to ensure safe and efficient take-off. The incorporation of high lift devices and hydrofoils can significantly improve take-off capabilities, especially in varying sea conditions. The power estimations underscored the significant power demands during take-off, necessitating efficient propulsion systems. The use of ducted underwater propellers or water jet propulsion units can help meet these demands, improving take-off performance and waterborne maneuverability.

The subsequent phases of the project will involve further refinement and optimization of the chosen design, the creation of digital twins and simulation environments, and the manufacturing and testing of prototypes. These steps are crucial for validating the design and ensuring that the WIG-craft can perform effectively in real-world environments.

Acknowledgements. This work was supported by the European Union's Horizon Europe research and innovation program under grant agreement No. 101096487, as part of the Airship project. The content reflects only the authors' views, and the European Union is not responsible for any use that may be made of the information it contains.

References

1. Alday Gonzalez, M. F. (2022). *Renewable Coarse Resource Assessment for the European Region.* Retrieved from https://euscores.eu/wp-content/uploads/2023/01/D6.2_CoarseAssessment-V2-TUDGL-20-12-2022-2.pdf
2. ANSYS, Inc. (2022). *Ansys Fluent User's Guide.* Retrieved 2023, from Ansys: https://ansyshelp.ansys.com/Views/Secured/corp/v221/en/flu_ug/flu_ug.html#flu_ug
3. ATR Aircraft. (2023). *ATR 72-600F Freighter Aircraft.* Retrieved 2023, from ATR Aircraft: https://www.atraircraft.com/our-aircraft/atr-72-600f-freighter/
4. Babikian, R. L. (2002). The historical fuel efficiency characteristics of regional aircraft from technological, operational, and cost perspectives. *Journal of Air Transport Management, 8*(6).
5. Baddoo, P. J. (2020). Exact solutions for ground effect. *Journal of Fluid Mechanics, 891.*
6. Chirosca, A.-M. (2022). Characteristics of the wind and wave climate along the European seas focusing on the Main maritime routes. *Journal of Marine Science and Engineering, 10*(1).
7. Dassaut Aviation. (2023). *Dassault Falcon 2000LXS.* Retrieved 2023, from Dassault Falcon: https://www.dassaultfalcon.com/businessjets/falcon-2000lxs/
8. Della Vecchia, P. (2014). Aerodynamic guidelines in the design and optimization of new regional turboprop aircraft. *Aerospace Science and Technology, 38.*
9. Gloudemans, R. A. (2022). Open vehicle sketch pad: An open source parametric geometry and analysis tool for conceptual Aircraft design. In *AIAA SCITECH 2022 forum.* American Institute of Aeronautics and Astronautics.
10. Graves, S. S. (1999). *Investigation of a Technique for Measuring Dynamic Ground Effect in a Subsonic Wind Tunnel.* NASA.
11. Gudmundsson, S. (2022). *General Aviation Aircraft design: Applied methods and procedures* (2nd ed.). Elsevier Science & Technology.
12. Gulfstream Aerospace. (2023). *Gulfsream Aerospace.* Retrieved 2023, from Gulfstream G280: https://www.gulfstream.com/en/aircraft/gulfstream-g280/
13. Kaario, T. (1935). *Finland Patent No. 18630/1935.*
14. Laine, S. H. (2006). *Lentokoneen aerodynamiikka ja lentomekaniikka.* WSOY.
15. Lippisch, A. M. (1964). *United States of America Patent No. US3190582A.*
16. Otsason, R., & H. O.-P. (2024). Business opportunities for a ground effect vehicle-case of Canary Islands. *Transport and Telecommunication.*
17. Phillips, W. F. (2013). Lifting-line predictions for induced drag and lift in ground effect. In *31st AIAA applied aerodynamics conference.* American Institute of Aeronautics and Astronautics.
18. Raymer, D. P. (1992). *Aircraft design: A conceptual approach.* American Institute of Aeronautics and Astronautics, Inc.
19. Richards, R. (2013). The "SeaStryder" Hydroplaning Floatwing Aircraft. *AERO 13 Conference.*
20. Rozhdestvensky, K. V. (2006). Wing-in-ground effect vehicles. *Progress in Aerospace Sciences, 42*(3).
21. Sommer, G. S. (1988). Ekranoplan: The Soviet Sea Monster. *Proceedings of U.S. Naval Institute, 114*(10).
22. Yun, L. B. (2010). *WIG craft and Ekranoplan: Ground effect craft technology.* Springer.

Innovative Air Mobility—Key Considerations for Rural Logistics Operations

Vadim Kramar[1](✉), Niina Mallat[2], Seppo Yrjölä[3], Virpi Oksman[4], and Minna Kulju[4]

[1] VTT Technical Research Centre of Finland, Oulu, Finland
vadim.kramar@vtt.fi
[2] Aalto University, Helsinki, Finland
niina.mallat@aalto.com
[3] Nokia, Oulu, Finland
seppo.yrjola@nokia.com
[4] VTT Technical Research Centre of Finland, Tampere, Finland
{virpi.oksman,minna.kulju}@vtt.fi

Abstract. This paper aims to revise Innovative Air Mobility from the perspective of rural logistics operations and to identify key considerations as an initial step in examining that specific aspect of operations. Reviewing and identifying stakeholders is a crucial part of this process because understanding them aids in initial planning, developing business models, and engaging with those stakeholders. The work does not intend to list rural logistics operations or justify their needs, as many publications cover that topic. Instead, it provides a holistic overview designed to help understand the entire knowledge domain based on available official documents, project deliverables and peer-reviewed studies. The primary limitation of this work is its focus on European regulations and the deliverables of research projects. Consequently, the paper outlines five key focus areas, highlights specific characteristics relevant to rural logistics operations, and provides a comprehensive list of stakeholders relevant to this context.

Keywords: Rural air mobility · Urban air mobility · Advanced air mobility · Innovative air mobility · U-space · UAS · eVTOL · VCA

1 Introduction

Unmanned Aircraft Systems (UAS) [1] and Vertical Take-off and Landing (VTOL) Capable Aircraft (VCA) [2] Technologies and applications are progressing swiftly, enabling UAS and VCA to become powerful solutions for various industries and application areas. In some innovative aerial services such as inspections, monitoring and mapping, UAS are gradually becoming a helpful tool of choice to replace dull, dirty or dangerous missions and more expensive traditional aircraft operations. At the same time, other areas, such as delivery, UAS and VCA, have the potential for significant market disruption. However, the extent of UAS, and particularly VCA, utilisation and market penetration, is still uncertain. Furthermore, over the past few years in Europe, the commercial applications of UAS have been significantly impacted by regulations. Since 2019, European

M. Laakkonen (ed.), *DroneAI for Resilient Nature*,
https://doi.org/10.1007/978-3-032-19008-6_13

regulatory frameworks have established requirements for the operations and technologies used in the UAS and VCA to be introduced into the European market. At the same time, the regulation introduced several new stakeholders relevant to the implementation of and operations within the U-space.

To match European regulations, U-space may be termed as a complex system designed to warrant the safe operation of any future aircraft, including UAS and VCA. To carry out these undertakings, European Union Member States are to assign specific geographical regions, known as U-space airspaces. This allocation is designed to support a range of U-space services that rely on digital technologies and automated functionalities. The primary objective of these services is to ascertain a secure, efficient and safe means of accessing U-space airspace [3]. The U-space concept is somewhat broader than the globally recognised concept of UAS Traffic Management (UTM), as it encompasses procedures and infrastructural considerations, the potential for sharing the same airspace with human-crewed aircraft operations, and convergence with Air Traffic Management (ATM), thereby dealing with a broader range of stakeholders.

The development and commercialisation of a competitive and sustainable end-to-end U-space system will essentially be a business model problem, stemming from innovation and transaction platforms and ecosystems that cover business, technology and policy perspectives. The earlier vision of the U-space was strongly associated with the concept of Urban Air Mobility [4], which in its turn implied not only urban but also suburban, peri-urban and inter-urban operations [5]. The European Drone Strategy 2.0 [6] broadened that vision by introducing the Innovative Air Mobility (IAM) concept and Innovative Aerial Services (IAS) (see Fig. 1).

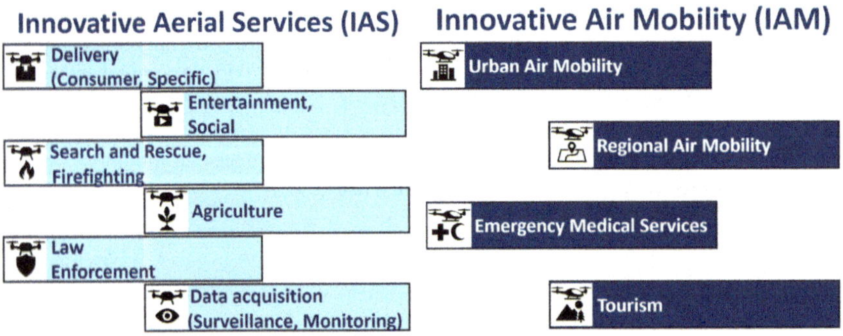

Fig. 1 Innovative aerial services and innovative air mobility. Adopted from EASA [7]

The non-EASA countries are developing similar concepts. For example, in the USA, the UTM concept proposed by the Federal Aviation Administration (FAA) is not a restricted airspace like the U-space in Europe, and is rather relevant to the operations of small UAS [8]. Similarly, the UAM concept primarily focuses on the operations of larger UAS and VCA carrying heavy cargo operations and transporting people [9]. These developments are well supported by the National Aeronautics and Space Administration (NASA). Around the globe, various countries are adopting either the EASA approach or the FAA/NASA approach, or exploring the benefits of both.

According to several sources, UAM is a subset of Advanced Aerial Mobility [10], also known by the more widely used term Advanced Air Mobility [9, 11], both of which share the same acronym, AAM. According to NASA, the concept of AAM can be traced back to the beginning of the XXI century [11], which makes it older than the concept of IAM introduced in 2022 [6]. To explore the potential of AAM, harmonise its development at the global level, and address the United Nations Sustainable Development Goals (SDGs), the International Civil Aviation Organisation (ICAO) has established the AAM Study Group [12].

To assess the future potential of UAS and VCA solutions, it is crucial to identify the key considerations and all relevant stakeholders within the broad business domain, encompassing its multidisciplinary aspects. Interestingly, all IAS and IAM operations examples may occur in rural areas. It is unclear, however, how rural operations may differ from urban operations, especially when those intersect or overlap in the same airspace. The regulatory changes reshape and extend the stakeholder map of the UAS and VCA industry. However, the current literature in the field lacks a comprehensive view of the particularities of rural operations, relevant stakeholders and their expectations of business value, as well as their position within the ecosystem.

This paper aims to fill that gap by presenting the results of a rigorous literature review of relevant and multiple regulatory and policy documents in the field, which currently provide the most comprehensive view of the different stakeholders involved. Based on the literature, a holistic view of the IAM elements is explored, and stakeholder roles are identified. The critical implications of the findings for IAM operations, with the attempt to recognise more specific details relevant to rural logistics operations that should be addressed for the UAS and VCA businesses to proliferate in the future, are established.

Research Question: *What are the key Innovative Air Mobility considerations for Rural Logistics Operations*?

A literature review was deemed the most suitable methodological approach for the study. Many search queries with different combinations of keywords, such as U-space, UAM, IAM, AAM, operations, urban and logistics, were executed using the Scopus and Google Scholar databases of scientific publications. Several 100 peer-reviewed publications were checked, including those focusing on general IAM concepts.

No underlying theory was recognised as applicable, but it has been found that the stakeholder-associated matters prevail in most publications, even though they are not thoroughly analysed. Indeed, the impact of stakeholders is crucial because aviation is one of the most regulated domains. Thus, the applicability of platform frameworks can be explored from both engineering and business perspectives [13]. In this dualistic tradition, the business has viewed platforms as a supply and demand of multi-sided markets, while from the engineering perspective, the focus has been on the modularity of technology and enabled innovation complementarities [14, 15]. Both approaches agree on a high-level platform structure comprising components and interfaces that form a networked system spanning the core actors of the platform. The research can lay the groundwork for subsequent stages, built upon and intertwined with the platform research [16].

A thorough examination of all relevant European regulatory documents and existing amendments was conducted. The documents were searched on the EUR-Lex portal [17], EASA [18], EUROCONTROL [19], SESAR JU [20] and JARUS [21] websites.

Deliverables of European projects, funded through various funding instruments provided by European Framework Programmes, such as Horizon 2020 and Horizon Europe, the SESAR Joint Undertaking, and national funding programmes of European countries, as well as regulatory documents, were also reviewed. The grey literature review was conducted primarily in accordance with the guidelines for working with the Grey Literature in Systematic Reviews for Management and Organisational Studies [22] and the AACODS Checklist [23], which evaluates grey sources based on authority, accuracy, coverage, objectivity, date and significance.

Following the introduction, the paper provides an overview of the key considerations relevant to IAM, emphasising the specific characteristics of rural logistics operations. In Sect. 3, an overview of relevant stakeholders is provided. Discussion and concluding remarks are given in Sect. 4.

2 Rural Logistics Operations

Both concepts, IAM and AAM, consider rural UAS and VCA operations, or rural air mobility [6, 9, 10, 24]. Essentially, inter-urban operations and even regional air mobility are projected to occur over rural areas. Nevertheless, due to the high expectations for UAM justified by higher business demand and more developed supportive infrastructure, the primary development of the IAM domain happens in urban areas. Still, due to the infancy of the entire domain, there is a lack of scientific publications about IAM. However, the anticipated similarities between urban and rural logistics operations allow for the identification and research of several essential considerations. Those are presented in the following subsections.

2.1 Sustainability

The concept of sustainability in relation to the IAM is somewhat challenging. There are no directly relevant Key Performance Indicators (KPIs) or measures to quantify such sustainability, which keeps it among the abstract or philosophical notions. According to some sources [25], in 1987, the United Nations (UN) Brundtland Commission defined sustainability as "meeting the needs of the present without compromising the ability of future generations to meet their own needs." Such a definition does not assume tangible measures.

In 2015, the UN presented the 2030 Agenda for Sustainable Development blueprint, which listed 17 Sustainable Development Goals (SDGs) [26]. The European Union (EU) acknowledges and follows the SDGs since sustainable development is one of the fundamental principles of the Treaty on the European Union and a priority objective for the EU's internal and external policies [27]. Among the ongoing actions are the voluntary reviews of SDG implementations, which include monitoring and reporting progress in an EU context, running the KnowSDGs Platform [28] and engaging civil society, the private sector and other stakeholders. Several projections from the SDGs to the IAM operations may be recognised.

Regarding Goal 3, to ensure healthy lives and promote well-being for all at all ages—mitigating pandemic cases, a more efficient implementation of delivery on demand, and

Emergency Medical Services (EMS) available in rural areas would have a positive impact [29–31].

Regarding Goal 4, to ensure inclusive and equitable quality education and promote lifelong learning opportunities for all—the IAM domain, due to the infancy of its development state, is expected to require a lot of educated professionals, qualified and certified, as it is typically necessary in aviation [8, 31–33].

Regarding Goal 9, to build resilient infrastructure, promote inclusive and sustainable industrialisation, and foster innovation—establishing quality, reliable, sustainable and resilient infrastructure to support economic development and human well-being, and increase passenger and freight volumes would be beneficial [8, 31, 34, 35].

Regarding Goal 11, to make cities and human settlements inclusive, safe, resilient and sustainable—rural logistics operations may contribute to the integrated and sustainable human settlement planning and essentially support positive economic, social and environmental links between urban, peri-urban and rural areas [10, 31, 36].

Regarding Goal 13, to take urgent action to combat climate change and its impacts—the environmental considerations that include CO_2, visual and noise pollution, are among the most important factors impacting a social acceptance of UAS and VCA operations [37, 38].

Targeting sustainability in Europe, the EC lists several areas of action towards the European industry [39]. Among those action areas are a circular economy, a competitive climate-neutral economy and sectoral initiatives. One of the sectoral initiatives is the strategy for a sustainable built environment—to achieve a climate-neutral, intelligent and resilient built environment for all EU citizens [40].

The EC introduced a new Circular Economy Action Plan for a cleaner and more competitive Europe [12] in 2020. It lists seven key product value chains, including electronics and ICT, batteries and vehicles, construction and buildings, that may be considered relevant to IAM technology, enabling aircraft design and ground infrastructure. At least, the Batteries and vehicles, and Construction and buildings value chains are explicitly mentioned in the EU Sustainable and Smart Mobility Strategy introduced by the European Commission in December 2020 [41], along with the reference to the Circular Economy Action Plan.

As the analysed literature shows, there is no difference between urban and rural operations, as the former implies the latter, and no more specific considerations are observed while fostering economic development. Improving the quality of life and addressing environmental concerns are also vital. These considerations are directly connected to social acceptance. Due to their complexity and modular nature, IAM solutions are expected to be developed in ecosystems centred around digital platforms [42], tailored to more than one stakeholder group, and having various impacts on sustainability.

2.2 Regulatory and Governance

To highlight the numerous regulatory and policy documents that collectively shape the future of IAM operations, this chapter provides a brief introduction to the most critical of these documents.

The primary regulatory frameworks affecting the IAS and IAM are the so-called EASA Basic Regulation, the UAS regulatory framework, and the U-space regulatory framework [43–45]. The Basic Regulation is Regulation 2018/1139 that sets the common rules in the field of civil aviation and for establishing the European Union Aviation Safety Agency [46]. The UAS regulatory framework includes Delegated Regulation 2019/945 [1], describing how the UAS can be offered in the EU market, and Implementing Regulation 2019/947 [47], describing the operations of UAS in open, specific and certified categories. The regulatory framework of the U-space includes Implementing Regulation 2021/664 [3], Implementing Regulation 2021/665 [48], and Implementing Regulation 2021/666 [49]. Regulation 2021/664 provides vital definitions, lists crucial stakeholders, and sets the demand for essential U-space requirements. Regulation 2021/665 amends the Implementing Regulation 2017/373 [50] concerning the standardised prerequisites for air traffic management (ATM), air navigation service (ANS) providers, and other functionalities of the ATM network within the allocated U-Space airspace within controlled airspace. Regulation 2021/666 amends Regulation 923/2012 [51] on the Standardised European Rules of the Air concerning the necessities for manned or crewed aviation operations in U-Space airspace.

The European Commission (EC) published several amendments to the UAS regulatory framework. Those include Delegated Regulation 2020/1058 [52] introducing two new UAS classes, Implementing Regulation 2020/639 [53] concerning standard scenarios for beyond the visual line of sight (BVLOS) operations, Implementing Regulation 2020/746 [54] deferring implementation deadlines for selected measures in light of the COVID-19 pandemic, Implementing Regulation 2021/1166 [55] concerning the postponement of the implementation date for standard scenarios applicable to operations conducted within line of sight or BVLOS and Implementing Regulation 2022/425 [56] on the extension of transition dates for the utilisation of specific UAS within the 'open' category, as well as the designated timeframe for implementing standard scenarios for operations conducted within line of sight or BVLOS. Also, Implementing Regulation 2023/203 [57], which prescribes guidelines for utilising Regulation 2018/1139 [46] of the European Parliament and of the Council regarding requirements for managing information security risks was published.

The presented in the autumn of 2022 Drone Strategy 2.0, is accompanied by the Commission Staff Working Document EU Drone Sector state of play [58].

In the spring of 2024, the EC published Implementing Regulation 2024/1111 [2], which lays the ground for the operations of a small category VCA.

The European Union Aviation Safety Agency (EASA) [18] published several Acceptable Means of Compliance (AMC) and Guidance Material (GM), clarifying the regulatory frameworks. To help understand the complex statements and interdependencies between existing UAS-relevant regulations, the EASA merged all regulatory documents into one and published them as Easy Access Rules. Such combined documents are updated periodically to comply with the most recent regulations, amendments and AMC/MGs.

The U-Space Concept of Operations (ConOps) is not a regulation, but a significant U-space community effort to provide technical and organisational details on how operations should occur within a U-space, including system characteristics of stakeholder interactions. The most recent version, ConOps edition 4.0, was released in 2023 [35], defined the U-space concept, essential elements, capabilities and services to enable

UAM. The significance of several editions of the European U-space ConOps documents is not only because some of the key European experts contributed to their development, but also because they wrapped up studies of many European projects researching and developing U-space and UAM, and obtained inputs from several rounds of consultancies and workshops with the European and international stakeholders active in the field.

Governance within this section refers to implementing regional and national regulatory demands and developing and implementing visions, strategies, processes and policies. That includes certification requirements and processes that stakeholders of rural logistics operations must be aware of. There is an ongoing debate about power relationships, gatekeeper roles and governance on a platform, particularly related to data and other resources. In the emerging IAM era, businesses can access and utilise a broader scope of resources, which calls for a comprehensive approach to creating, sharing and capturing value [59].

2.3 Aircraft Technologies

These considerations include airframe designs and core, as well as enabling and operating technologies. Among those are powertrain technologies; power source and battery technologies; materials and mechanical technologies; avionics and mechatronics; electronics, electricity and sensor technologies; computer, communication and software technologies; operational data analytics, information systems, geoinformatics and navigation; operational Artificial Intelligence (AI), machine learning and automation technologies; Human-Computer-Interaction (HCI); computer vision, flight control and operational safety technologies; data and cyber security, privacy and maintenance technologies; operational logistics, autonomy levels and integration in supply chain and business processes systems [60–66].

Depending on the business needs and the delivery object, such as cargo or passenger(s), the operational requirements determine the set of technology requirements for the aircraft. Those may include the airframe type, power source, Maximum Take-off Mass (MTOL), payload mass, endurance or operating range, etc. Additionally, legal requirements may be imposed to achieve the required level of safety, which may imply certification of the aircraft, its components and/or the operations. For example, UAS may require a Specific Assurance and Integrity Level (SAIL), and the necessary volume of operations may be feasible only when performed by a certified UAS operator.

In addition to UAS, also known as drones, and VCA, other aircraft types may also operate within the IAM concept. Those include such fixed-wing aircraft as Conventional Take-off and Landing (CTOL), also known as Horizontal Take-off and Landing (HTOL), and Short Take-off and Landing (STOL) aircraft that require shorter runways. Emerging Technology Aircraft (ETA) are aircraft that are not covered by existing categories in the current ICAO Annex 16 Volume 1 Certification Procedures [67].

Although rural operations may have longer operational ranges than urban operations, the literature review does not reveal any additional requirements [10]. This is also the case with regional air mobility, where operational ranges are estimated to be several 100 km. As a vertical industry, aircraft technologies may be characterised as incumbent-born platforms. Those are grounded on the core competencies and products that provide infrastructure and services to their traditional customer segments [68]. These upstream-side platforms are vertically oriented supply-side platforms [42], leveraging a modular architecture with interfaces that permit the introduction of complementarities, such as technological capabilities, enabling rural logistics operations.

2.4 Supporting Infrastructure

Availability and maturity of ground infrastructure, including vertiports, emergency landing, maintenance, power and water supply, passenger/cargo and waste management, are among the crucial considerations of IAM operations [36, 64, 65, 69–71].

Tailored to ground infrastructure, additional services such as weather stations, forecasting systems and terrestrial or satellite telecommunications may be required. ICT networks and services, including cloud services and data lakes, might be needed to collect and process data. Those systems should be protected with data and cybersecurity means.

So far, the EASA Technical Design Specifications for Vertiport is still a prototype [72], while the recently published ISO 5491:2023 standard for vertiport infrastructure and equipment [73] had yet to gain popularity in Europe.

The lack of ground infrastructure, additional services and systems may set a barrier at the early stages of adopting rural logistics operations [10] and require more intensive business development operations and higher investment. For example, rural areas may not be well covered with mobile network services for the entire altitude of expected operations [74] and it is a good question whether the expected volume of operations will bring enough revenue to cover capital expenditures (CapEx) and justify the operator's expected return on investment (ROI).

Supportive infrastructure operations are typically based on horizontal business models and form an innovation-driven platform that is born adjacent, focusing on serving users on the downstream side [68]. The technology enablers and capabilities, particularly those associated with communications, cloud, data and Artificial Intelligence (AI), enable expansion to demand-side phantomization and multi-sided platforms with transaction focus [42].

2.5 Airspace Integration

Airspace integration is a very complex topic that includes airspace design, flight rules, levels of automation, digital information and data management, interoperability and even convergence of ATM and U-space [8, 35, 75–77]. The majority of considerations are sophisticated matters ranging well beyond the scope of this work, while some research reports admit that rural operations are less challenging and complex than urban [10].

At the level of planning and organising rural logistics operations, critical considerations are what the aircraft is, in what airspace class, as listed in ICAO Annex 11 [78] It operates, and whether it is required to operate within the U-space or outside the U-space restricted zone, with or without the support of U-space services. Those moments will require additional consideration, but unfortunately, current regulations do not entirely cover all of them in segregated airspace volumes.

Multi-sided platforms as business ecosystems utilise without owning the hardware, software, data and service assets and assist, control and govern their exchange across organisations [42]. Regulated and standardised IAM systems with open interfaces to novel digital services and the automation of functions can be pro-competitive by lowering the market entry barrier, contributing to complementarities and profiting from innovations.

3 Stakeholder Landscape

This section provides a comprehensive overview of the U-space and the various forms of air mobility stakeholders developed based on the work conducted in the AiRMOUR project [79, 80]. The stakeholders were identified through an extensive literature review of various regulatory and policy documents, European project deliverables, and peer-reviewed publications. In cases of competing definitions, priority was given to the document's rank and its publication date. Recently adopted regulatory documents had the highest priority.

The list of stakeholders and their roles is depicted in Table 1. Table 1 consists of four columns. The first column lists the formal name of the stakeholder, as defined in a credible source. The second column is the description obtained from the source. The third column includes references to credible sources that define the stakeholder. The final column provides information about the group to which the stakeholder belongs. A summary of the groups is listed in Table 2, and an explanation of the grouping, tailored to rural logistics operations, is provided below.

The first group of stakeholders is the Beneficiaries. Those are such demand-side stakeholders as air taxi passengers, consumers, end users, operation customers, VCA passengers, UAS delivery clients and generalised UAS delivery customers. Like the general public and passenger, end-user and client stakeholders of the demand side, assemblies of people and uninvolved persons are also generalised as part of this stakeholder group. Whenever relevant, the stakeholders of this group are to be served by rural logistics operations.

The Actors of IAM form the second group. Those stakeholders are similar to the actors of other transportation modalities, such as roads, rails and inland waterways; however, unlike these, they are directly involved in air or aerial operations on a personal level. The organisers of rural logistics operations have to be aware of actors that may share the same airspace, such as paragliders and parachutists, and whenever appropriate, may engage some actors, such as chief pilots, PICs, crew members, airspace observers and others. Legislation sets the rules of engagement.

The third group is Service providers and operators. Stakeholders belonging to this group facilitate and ensure aircraft operations, thereby providing a foundation for a supply-side approach. Such stakeholders include commercial or general aviation operators, airport or vertiport operators, UAS/VCA operators, other aircraft operators, AISPs, ANSPs, CNS SPs, CISPs, SDSPs, USSs and USSPs, among others. The organisers of

Table 1 IAM stakeholder landscape

Stakeholder	Description	Source	Group
Accident investigation body	Conducts an accident investigation	[81, 82]	6
Aerodrome operator	Operates one or more aerodromes	[35, 46]	3
Aeronautical information service (AIS) provider (AISP)	Provides aeronautical information and data for air navigation	[35]	3
Air delivery booking service provider	Delivers services allowing booking a delivery by air	[35]	3
Air taxi booking service provider	Delivers services allowing booking of an air taxi	[35]	3
Air taxi passenger	A passenger of an air taxi	[35]	1
Air traffic management (ATM)/air navigation services (ANS) provider (ANSP)	Provides air navigation services in controlled airspace and close to it	[83]	3
Air traffic service (ATS) provider	Provides air traffic services to airspace users	[35]	3
Aircraft operator	Operates or proposes to operate one or more aircraft	[46]	3
Airport operator	Operates one or more aerodromes	[84]	3
Airspace observer	Assists the remote pilot visually, detects hazards and provides alerts	[46]	2
Airspace user (other than UAS/UAM)	Operates or owns the aircraft	[35, 84]	1
Assemblies of people	Encircles dense crowds of people that cannot easily move	[1]	1
Authorized representative	Operates on behalf of a manufacturer	[1]	4
Authority for safety and security	Prepares and supervises the operations of law enforcement	[35]	6
Aviation user	Uses the airspace as a manned aircraft	[83]	2
Border control authority	Carries out border checks	[1, 85]	6
Cargo-hub operator	Operates or proposes to operate one or more cargo vertiports	[35]	3
Civil aviation authority	Encircles national or local aviation authorities	[83]	6
CNS Service provider (CNS SP)	Provides services for communication, navigation or surveillance	[35]	3
Common information service provider (CISP)	Provides services according to the U-space regulatory framework	[35]	3

(*continued*)

Table 1 (*continued*)

Stakeholder	Description	Source	Group
Competent authority	Performs tasks as in article 18 of 2019/947	[47]	6
Conformity assessment body	Performs conformity assessment activities	[1]	6
Consumer	Acts for purposes not associated with own trade, business, craft or profession	[1, 86]	1
Distributor	Renders a particular product accessible within the market	[1]	4
Drone association	Represents and assists stakeholders	[83]	4
Economic operator	Carries out economic operations associated with UAS	[1]	4
Emergency responder	Prepares and executes emergency operations	[35]	6
End user	Consumes a UAS-delivered service	[43]	1
Equipment manufacturer	Develops hardware solutions for U-space	[83]	4
European commission	Shapes the EU'S overall strategy, facilitates UAS-associated development, proposes policies and regulations, forms required institutions, controls and supervises relevant activities	[83, 87]	7
European union aviation safety agency (EASA)	Ensures the highest common level of safety and environmental protection for EU citizens. Supports the member states and the EU Council. Works with international aviation organizations and regulators	[18, 83]	7
European defence agency (EDA)	Develops policies and supports the member states and the EU Council regarding defence capabilities	[83, 88]	7
European organization for civil aviation equipment (EUROCAE)	Develops standards for European civil aviation	[89]	5
European organization for the Safety of air navigation (EUROCONTROL)	Supports member states and civil and military stakeholders to achieve safe, efficient, cost-effective and environmentally friendly aviation operations	[19, 83]	7

(*continued*)

Table 1 (*continued*)

Stakeholder	Description	Source	Group
General public	Encircles involved users and the uninvolved public	[35]	1
Global UTM association (GUTMA)	Represents organisations involved in innovative drone services and smart mobility solutions to keep drone operations safe and efficient at all altitudes.	[90]	5
Government entity	See the accident investigation body and competent authority	[81]	6
Infrastructure service provider (SP)	Provides infrastructure	[35]	3
Innovative air mobility (IAM) operator	Performs IAM operations and carries out accountability for them	[6]	3
Importer	Situated within the European Union entity that engages in the transportation and trade of a product from a country outside the EU to be sold within the EU market.	[1]	4
Insurance company	Collects statistics about rates of drone accidents in U-space and offers insurance policies	[83]	3
Insurance data service provider	Provides insurance-related supplemental data	[83]	3
Joint authorities for rulemaking on unmanned systems (JARUS)	Develops safety, technical and operational requirements for all qualities linked to the UAS operations	[21, 83]	5
Local and specific authorities	Defines specific operating procedures and rules and facilitates execution of those	[35]	6
Manned aircraft manufacturer	Produces manned aircraft that is compliant with regulatory demand	[83]	4
Manufacturer	Produces a product and markets it	[1]	4
Market surveillance authority	Carries out market surveillance on its territory	[1]	6
Model aircraft club or association	Resides within the EU and encircles members conducting leisure or sporting activities using UAS	[47]	4

(*continued*)

Table 1 (*continued*)

Stakeholder	Description	Source	Group
Model aircraft operator	Operates a model aircraft, which is considered as UAS as stated in 2018/1139	[46]	3
National accreditation body	Responsible for carrying out accreditation processes based on the authority derived from the EU member states	[91]	6
National competent authority	Encircles organizations that an individual member state has appointed and possesses the requisite authorities and assigned duties within their respective areas of expertise as per regulations 2018/1139.	[46]	6
National supervisory authority	Assumes the responsibility of ensuring compliance with aviation legislation, safeguarding all aircraft's safe and secure operations; actively strives to promote the diminution of environmental consequences and proactively anticipates potential hurdles in the deployment process	[83]	6
Notified body	Receives notifications in order to provide conformity assessment services	[1]	6
Notifying authority	Provides notification to and monitoring of notified bodies	[1]	6
Operation customer	Consumes the result of the UAS operation and perhaps authorises the mission	[83]	1
Qualified entity	Performs certification or oversight tasks as in 20,218/1139	[46]	4
Remote pilot	Performs the UA flight, and carries out the associated tasks and responsibilities as in 2019/945	[1]	2
Single European sky ATM research 3 joint undertaking (SESAR 3 JU)	Accelerates the implementation of the digital European sky strategy through research activities and innovation	[20, 83]	7
Specialised press	Conducts communication and dissemination activities	[83]	4

(*continued*)

Table 1 (*continued*)

Stakeholder	Description	Source	Group
Supplemental data service provider (SDSP)	Offers access to supplemental data sources that effectively support the U-space services	[35]	3
Technology vendor	Delivers technology solutions for U-space	[43]	4
Training organisation	Carries out pilot and operator training	[83]	4
UAM operator	Performs UAM operations and carries out accountability for them	[35]	3
UAM passenger	Benefits from being transferred as a result of the UAM operation	[35]	1
UAS command unit operator	Holds a command unit operator certificate (CUOC) and ensures that the command unit and any installed component adhere to the certified design, specifications and performance standards and thus support safe operation of the unmanned aircraft.	[35]	3
UAS delivery client	Physical person who benefits from the UAS delivery service	[35]	1
UAS delivery customer	Legal entity that benefits from the UAS delivery service	[35]	1
UAS manufacturer	Produces UAS and/or relevant technology solutions and holds relevant certificates	[35]	4
UAS service provider	Delivers services that are provisioned with the help of UAS	[43]	3
UAS operator	Operates or intends to operate one or more UAS	[1]	3
Uninvolved person	Is not aware of or participates in the UAS operation	[47]	1
Universities, academic institutions and research projects	Conducts research, development and innovation activities and generates knowledge	[83]	4
U-space coordinator	Coordinates with other stakeholders on the U-space matters	[3, 35]	6
U-space service industry	Develops software for U-space	[83]	4

(*continued*)

Table 1 (*continued*)

Stakeholder	Description	Source	Group
U-space service provider (USSP)	Provides the U-space services as in 2021/664	[3, 35],	3
Vertiport operator	Operates or proposes to operate one or more vertiports	[72]	3

Table 2 Stakeholder groups

Group	Description
1	Beneficiaries
2	Actors
3	Service providers and operators
4	Other stakeholders relevant to IAM economic activities
5	Standards development organizations (SDOs) and industry associations
6	National, regional and local governing bodies and authorities
7	Multi-state governing and other crucial bodies and authorities

rural logistics operations must be well aware of these stakeholders and interact with them as required by regulations or justified by business needs.

The fourth group is other relevant to IAM economic activities stakeholders. This group of stakeholders is the most diverse, comprising both commercial and non-profit stakeholders who act in the interests of their economic activities, as well as bystanders. The group includes manufacturers and vendors, academia and research institutions, NGOs, environmental and sustainability advocates, specialised press, interest groups and clubs, local communities and public engagement groups. The organisers of rural logistics operations may interact with these stakeholders due to business needs and, notably, due to sustainability considerations. The latter implies both acquiring sustainable aircraft, equipment and services, as well as promoting sustainable actions, which may help lay the groundwork for social acceptance.

The fifth group is Standards Development Organisations (SDOs) and industry associations. The organisers of rural logistics operations are unlikely to interact with this stakeholder group. Still, following the activities of those stakeholders may help with a better understanding of the IAM domain.

The sixth group is National, regional and local governing bodies and authorities. It is vital to follow these stakeholders to be aware of and compliant with regulatory and policy demands, and to interact with them to achieve compliance, as well as whenever required, when planning or implementing rural logistics operations. This category includes national security, defence and law enforcement stakeholders, interactions with which are likely outside the scope unless justified.

The final, seventh group is multi-state governing and other crucial bodies and authorities. Organisations of rural logistics operations are unlikely to interact directly with this group. Similarly to Group Six, following the activities of these stakeholders may help with a better understanding of not only the IAM domain but also develop a broader sense of future strategies and policies.

Among the best practices are early engagement with different stakeholders, including regulatory authorities and business partners, and close follow-up of the development of national regulatory and authority entities by all stakeholders, because their actions, plans and schedules [92].

4 Discussion and Conclusions

Some research reports show [10] that developing a rural cargo delivery system could significantly enhance the aerial mobility market. There is a need to cover large distances efficiently, with major air carriers exploring the use of UAS and VCA for rural deliveries. Combining these with traditional delivery trucks can improve efficiency and reduce costs, while also helping to gauge societal acceptance of these aircraft. Rural areas present ideal conditions for UAS and VCA operations due to lower population density, minimal air traffic, fewer obstacles and the need for long-distance delivery with a small logistics footprint. This supports local communities and serves as a crucial testing ground for future IAM operations. Although it is possible to anticipate that the IAM market in rural areas will not be clearly divided into rural and urban mobility, it will instead develop as a mixed state.

The other research demonstrates [93] that in certain areas, the negative environmental impact expected from UAS delivery in rural areas may be 13 times smaller than that in urban areas. Not tailored to any mobility, environmental benefits (11.3%) are confirmed by the other studies [94], along with the economic (49.3%) and societal (20.2%) benefits. At the same time, legal aspects (23.9%), ethical aspects (22.7%), physical safety (22.0%), social issues (12.8%), environmental interrelations (7.5%) and economic problems (6%) are not tailored to any mobility potential problems. Similarly, among the listed barriers are technical aspects (49.1%), legal issues (23.7%), lack of public acceptance (15.7%), economic barriers (6.6%) and infrastructural barriers (4.9%).

Ecosystem-focused platform business models, commonly found in software, cloud and e-commerce contexts, can emerge in the IAM segment, particularly in rural logistics operations. Related to the digitalised services, as a service (aaS) business models [95] are widely deployed, facilitating a variety of digital cloud-based service providers [96]. In these ecosystemic platform-based business models, which focus on the new complementary service co-creations, the openness of the business model [97] and the open business model innovation concept [98] that connects the ecosystem and its stakeholders was found essential.

Open innovation and interlinked complementarities raise concerns over the stakeholders' capabilities to profit from innovation due to positive externality spillovers related to technologies [16]. Our findings align with [99] in that the critical elements of a successful business model, providing scalability, replicability and sustainability, in addition to the innovations, functionalities and features of the core product, include several aspects of the multi-sided platform, a source of complementarities and platform governance. Furthermore, in emergent IAM operations, stakeholders should acknowledge the role of co-opetition, i.e., how to balance their collaborative and competitive actions associated with a platform between partners and complements, as well as between platforms [16].

This paper identified a need to combine the characteristics of innovation and transaction platforms to introduce the IAM ecosystem successfully. Innovation platforms create value by coping with constraints associated with technological innovation, while transaction platforms' value creation is related to the coordination of consumption problems. Furthermore, identified platform archetypes comprise interconnected, multilayered subsystems connected via standardised interfaces, providing assured interoperability. Stakeholder analysis emphasises the emerging role of multi-sided transaction platforms. Moreover, the value co-creation and value co-capture, as evident in the compilation of complementary stakeholders interconnected through standardised technologies and open-sourced assets, were clearly visible in the results.

The emergent IAM ecosystem, which includes new roles and actors, a multifaceted environment and systemic organisational configurations, will have the challenge of mastering the liability of newness. Value creation in the IAM platforms will depend highly on the developers of various complements. In contrast, the value proposition of an emergent ecosystem is grounded on the user-perceived quality of service and value delivered. An orchestrator, complementing developers, users and external actors, plays a vital role in maintaining the common ecosystem's legitimacy. External actors, including policymakers, regulatory authorities, research institutes, competition authorities, media and market researchers, play a crucial role in enabling, framing and challenging the emergence of the business ecosystem and the construction of legitimacy. Policymakers and regulatory bodies can facilitate the legitimation processes, especially via proactive and anticipatory regulation and standardisation that consider the IAM ecosystem and its role in society.

Because the entire IAM domain is still in its early infancy, there is a lack of comparative analysis of urban and rural mobility. Another reason for this is that the primary focus is currently on urban operations, where the expected volume of operations is sufficient to cover capital expenditures (CapEx) and operational expenses (OpEx).

Still, the authors believe that a combination of IAM operations in suburban and rural areas is the near-term development of the IAM domain, characterised by low density but higher experimental operations in rural areas. Additionally, the authors acknowledge that, except for the density of operations, the actual difference between urban and rural logistics operations may not be as significant as it appears. These movements still need to be verified.

Nevertheless, the authors identified five focus areas that form the entire IAM domain and recognised some of those moments that set apart rural logistics operations. Due to the limited outcomes from the literature review, this work requires continuation by applying a combination of research methods, including experimental ones.

Regarding the IAM stakeholder landscape, the authors compiled the most comprehensive list ever published and successfully grouped the stakeholders, despite their heterogeneous nature. The next step for that work would be to tailor stakeholder groups to the technology platforms and link them to the five focus areas discovered and described in this paper.

Acknowledgements. This research is conducted under the European Union's Horizon 2020 Research and Innovation Programme-funded AiRMOUR (grant agreement 101006601) project and the Business Finland-funded DROLO (grant agreement 41170/31/2020) and DROLO II (grant agreement 4002/31/2023) projects.

References

1. European Commission (2019, March). *Commission delegated regulation (EU) 2019/945 of 12 March 2019 on unmanned aircraft systems and on third-country operators of unmanned aircraft systems* (pp. 1–40). European Commission (Online). Available: http://data.europa.eu/eli/reg_del/2019/945/oj
2. European Commission (2024). *Commission implementing regulation (EU) 2024/1111 of 10 April 2024 amending regulation (EU) No 1178/2011, implementing regulation (EU) No 923/2012, regulation (EU) No 965/2012 and implementing regulation (EU) 2017/373, as regards the establishment of R*. Accessed: 01 December 2024 (Online). Available: https://eur-lex.europa.eu/eli/reg_impl/2024/1111/oj
3. European Commission (2021, April). *Commission implementing regulation (EU) 2021/664 of 22 April 2021 on a regulatory framework for the U-space* (p. C/2021/2671, 161–183) (Online). Available: http://data.europa.eu/eli/reg_impl/2021/664/oj
4. Hately, A., & CORUS-XUAM (2022). *U-space ConOps* (Online). Available: https://corus-xuam.eu/wp-content/uploads/2022/11/CORUS-XUAM-D4.1-delivered_3.10.pdf
5. EU Funding & Tenders Portal. *Towards sustainable urban air mobility MG-3-6-2020*. Accessed: 03 December 2019 (Online). Available: https://ec.europa.eu/info/funding-tenders/opportunities/portal/screen/opportunities/topic-details/mg-3-6-2020
6. European Commission (2022, November). *A drone Strategy 2.0 for a smart and sustainable unmanned aircraft eco-system in Europe* (pp. 1–25) (Online). Available: https://eur-lex.europa.eu/legal-content/EN/TXT/?uri=CELEX:52022DC0652
7. EASA. *Drones & air mobility basics explained*. Accessed: 16 August 2024 (Online). Available: https://www.easa.europa.eu/en/domains/drones-air-mobility/drones-air-mobility-landscape/basics-explained
8. FAA (2020). *Concept of operations v2.0. Unmanned aircraft systems (UAS) traffic management (UTM)*. https://doi.org/10.1201/9781003124689
9. Federal Aviation Admistration (2023). *Urban air mobility (UAM) concept of operations v2.0* (Online). Available: https://www.faa.gov/sites/faa.gov/files/UrbanAirMobility%28UAM%29ConceptofOperations2.0_1.pdf
10. *Committee on enhancing air mobility, advancing aerial mobility: A national blueprint* (2020). Committee on enhancing air mobility—A national blueprint. https://doi.org/10.17226/25646
11. Patterson, M. (2021). *Advanced air mobility (AAM): An overview and brief history*. NASA.
12. Abboud, D. *Navigating the intersection of advanced air mobility and sustainable development goals*. Uniting Aviation. Accessed: 30 October 30 (Online). Available: https://unitingaviation.com/news/capacity-efficiency/navigating-the-intersection-of-advanced-air-mobility-and-sustainable-development-goals/

13. Gawer, A. (2014). Bridging differing perspectives on technological platforms: Toward an integrative framework. *Research Policy, 43*(7), 1239–1249. https://doi.org/10.1016/j.respol.2014.03.006
14. Gaynor, M., & Bradner, S. (2001, July). Using Real Options to Value Modularity in Standards. *Knowledge Technology Policy, 14*, 41–66. https://doi.org/10.1007/s12130-001-1005-6
15. Schilling, M. (2000, April). Toward a general modular systems theory and its application to interfirm product modularity. *The Academy of Management Review, 25*. https://doi.org/10.5465/AMR.2000.3312918
16. Teece, D. J. (2018, October). Profiting from innovation in the digital economy: Enabling technologies, standards, and licensing models in the wireless world. *Research Policy, 47*(8), 1367–1387. https://doi.org/10.1016/J.RESPOL.2017.01.015
17. EU law—EUR-Lex. Accessed: 30 January 2023 (Online). Available: https://eur-lex.europa.eu/homepage.html
18. EASA|European Union Aviation Safety Agency. Accessed: 30 January 2023 (Online). Available: https://www.easa.europa.eu/en
19. EUROCONTROL|Supporting European Aviation. Accessed: 30 January 2023 (Online). Available: https://www.eurocontrol.int/
20. SESAR Joint Undertaking|Delivering the Digital European Sky. Accessed: 30 January 2023 (Online). Available: https://www.sesarju.eu/
21. Joint Authorities for Rulemaking on Unmanned Systems|JARUS. Accessed: 30 January 2023 (Online). Available: http://jarus-rpas.org/
22. Adams, R. J., Smart, P., & Huff, A. S. (2017). Shades of Grey: Guidelines for working with the Grey literature in systematic reviews for management and organizational studies. *International Journal of Management Reviews, 19*(4), 432–454. https://doi.org/10.1111/ijmr.12102
23. Tyndall, J. (2010). *The AACODS checklist is designed to enable evaluation and critical appraisal of grey literature.*
24. Andritsos, K., Scott, B. I., & Trimarchi, A. (2022). What is in a name: Defining key terms in urban air mobility. *Journal of Intelligent & Robotic Systems, 105*(4), 1–9. https://doi.org/10.1007/s10846-022-01694-1
25. United Nations. *Sustainability*. Accessed: 31 August 2023 (Online). Available: https://www.un.org/en/academic-impact/sustainability
26. United Nations. *Sustainable development goals|United Nations development programme.* Accessed: 18 April 2023 (Online). Available: https://www.undp.org/sustainable-development-goals
27. European Commission. *Sustainable development goals.* Accessed: 31 August 2023 (Online). Available: https://commission.europa.eu/strategy-and-policy/sustainable-development-goals_en
28. European Commission. *The knowSDGs platform.* Accessed: 31 August 2023 (Online). Available: https://knowsdgs.jrc.ec.europa.eu/
29. Johnson, A. M., Cunningham, C. J., Arnold, E., Rosamond, W. D., & Zègre-Hemsey, J. K. (2021). Impact of using drones in emergency medicine: What does the future hold? *Open Access Emergency Medicine, 13*, 487–498. https://doi.org/10.2147/OAEM.S247020
30. Kramar, V. (2020). UAS (drone) in response to coronavirus. In *Proceedings of the 27th conference of FRUCT association* (pp. 90–100). IEEE Computer Society. https://doi.org/10.23919/FRUCT49677.2020.9211075
31. Cohen, A. P., Shaheen, S. A., & Farrar, E. M. (2021). Urban air mobility: History, ecosystem, market potential, and challenges. *IEEE Transactions on Intelligent Transportation Systems, 22*(9), 6074–6087. https://doi.org/10.1109/TITS.2021.3082767

32. Cohen, A., Shaheen, A., Francisco, S., & Cohen Susan Shaheen, A. (2023, March). *Advancing aerial mobility through technology, sustainability, and on-demand flight*. https://doi.org/10.7922/G2PN93Z8
33. Battiste, V., Dao, A. Q. V., Strybel, T. Z., Boudreau, A., & Wong, Y. K. (2016). Function allocation strategies for the unmanned aircraft system traffic management (UTM) system, and their impact on skills and training requirements for UTM operators. *IFAC-PapersOnLine, 49*(19), 42–47. https://doi.org/10.1016/j.ifacol.2016.10.459
34. Levitt, I., Phojanamongkolkij, N., Horn, A., Witzberger, K. (2023). *UAM airspace research roadmap Rev 2.0* (Online). Available: https://www.sti.nasa.gov/
35. SESAR 3 JU & CORUS-XUAM (2023). *U-space ConOps and Architecture* (4th Edn.) (Online). Available: https://www.sesarju.eu/sites/default/files/documents/reports/U-spaceCONOPS4thedition.pdf
36. Long, Q., Ma, J., Jiang, F., & Webster, C. J. (2023, September). Demand analysis in urban air mobility: A literature review. *Journal of Air Transport Management, 112*, 102436. https://doi.org/10.1016/J.JAIRTRAMAN.2023.102436
37. Rizzi, S. A., et al. (2020). *Urban air mobility noise: Current practice, gaps, and recommendations*. NASA.
38. Oksman, V., Kulju, M., & Kramar, V. Exploring the factors influencing public acceptance and intention to use automated drone delivery systems: A trust theory approach. In *2024 AIAA DATC/IEEE 43rd AIAA/IEEE digital avionics systems conference* (pp. 1–9). https://doi.org/10.1109/DASC62030.2024.10748821
39. European Commission. *Internal market, industry, entrepreneurship and SMEs—Industry—Sustainability*. Accessed: 31 August 2023 (Online). Available: https://single-market-economy.ec.europa.eu/industry/sustainability_en
40. European Commission. *Internal market, industry, entrepreneurship and SMEs—Industry—Buildings and construction*. Accessed: 31 August 2023 (Online). Available: https://single-market-economy.ec.europa.eu/industry/sustainability/buildings-and-construction_en
41. European Commission (2020). Sustainable and smart mobility strategy—Putting European transport on track for the future.
42. Teece, D. J., Pundziene, A., Heaton, S., & Vadi, M. (2022, August). Managing multi-sided platforms: Platform origins and go-to-market strategy. *California Management Review, 64*(4), 5–19. https://doi.org/10.1177/00081256221109961
43. Ojanen, P., & Yrjola, S. (2022). The Spectrum policy challenges of unmanned traffic management and U-space regulation—Stakeholder analysis view. In *2022 International conference on unmanned aircraft systems ICUAS* (pp. 714–723). https://doi.org/10.1109/ICUAS54217.2022.9836076
44. Kramar, V., Röning, J., Nikolakopoulos, G., & Tomasello, F. (2021, October). Urban air mobility overview-the European landscape. In *Proceedings of the 30th conference of FRUCT association, Oulu: 30th conference of open innovations association FRUCT* (pp. 99–106). https://doi.org/10.23919/FRUCT53335.2021.9599973
45. Barbano, M., & Costa, V. (2023). Implementing urban air mobility in a multi-level regulatory framework: Perspectives from the EU. In *2023 International conference on unmanned aircraft systems ICUAS 2023* (Vol. 2, No. 1, pp. 895–902). https://doi.org/10.1109/ICUAS57906.2023.10156184
46. European Parliament and Council (2018). *Regulation (EU) 2018/1139 of the European Parliament and of the Council of 4 July 2018 on common rules in the field of civil aviation and establishing a European Union aviation safety agency, and amending regulations (EC) No 2111/2005, (EC) No 1008/2008* (p. PE/2/2018/REV/1, 1–122) (Online). Available: http://data.europa.eu/eli/reg/2018/1139/oj

47. European Commission (2019, May). *Commission Implementing Regulation (EU) 2019/947 of 24 May 2019 on the rules and procedures for the operation of unmanned aircraft* (p. C/2019/3824, 45–71). European Commission (Online). Available: http://data.europa.eu/eli/reg_impl/2019/947/oj
48. European Commission (2021). *Commission implementing regulation (EU) 2021/665 of 22 April 2021 amending implementing regulation (EU) 2017/373 as regards requirements for providers of air traffic management/air navigation services and other air traffic management network functions in 2021* (p. C/2021/2672, 184–186) (Online). Available: http://data.europa.eu/eli/reg_impl/2021/665/oj
49. European Commission (2021). *Commission implementing regulation (EU) 2021/666 of 22 April 2021 amending regulation (EU) No 923/2012 as regards requirements for manned aviation operating in U-space airspace* (p. C/2021/2673, 187–188) (Online). Available: http://data.europa.eu/eli/reg_impl/2021/666/oj
50. European Commission (2017, March). *Commission implementing regulation (EU) 2017/373—of 1 March 2017—Laying down common requirements for providers of air traffic management/air navigation services and other air traffic management network functions and their oversight, repealing Regulation* (pp. 1–126) (Online). Available: http://data.europa.eu/eli/reg_impl/2017/373/oj
51. European Commission (2012). *Commission implementing regulation (EU) No 923/2012 of 26 September 2012 laying down the common rules of the air and operational provisions regarding services and procedures in air navigation and amending implementing regulation (EU) No 1035/2011 and regulation* (Vol. 20, No. 1, pp. 1–66) (Online). Available: http://data.europa.eu/eli/reg_impl/2012/923/oj
52. European Commission (2020, April). *Commission delegated regulation (EU) 2020/1058—of 27 April 2020—amending delegated regulation (EU) 2019/945 as regards the introduction of two new unmanned aircraft systems classes* (pp. 1–27) (Online). Available: http://data.europa.eu/eli/reg_del/2020/1058/oj
53. European Commission (2020, May). *Commission implementing regulation (EU) 2020/639 of 12 May 2020 amending implementing regulation (EU) 2019/947 as regards standard scenarios for operations executed in or beyond the visual line of sight* (pp. 1–31) (Online). Available: http://data.europa.eu/eli/reg_impl/2020/639/oj
54. European Commission (2020, June) *Commission implementing regulation (EU) 2020/746 of 4 June 2020 amending implementing regulation (EU) 2019/947 as regards postponing dates of application of certain measures in the context of the COVID-19 pandemic* (Vol. OJ L 176, pp. 13–14) (Online). Available: C/2020/3599.
55. European Commission (2021). *Commission implementing regulation (EU) 2021/1166 of 15 July 2021 amending implementing regulation (EU) 2019/947 as regards postponing the date of application for standard scenarios for operations executed in or beyond the visual line of sight* (pp. 49–50). Accessed: 20 August 2021 (Online). Available: http://data.europa.eu/eli/reg_impl/2021/1166/oj
56. European Commission (2022, March). *Commission implementing regulation (EU) 2022/425 of 14 March 2022 amending implementing regulation (EU) 2019/947 as regards postponing the transition dates for the use of certain unmanned aircraft systems in the 'open' category and the date of application* (pp. 20–21).
57. European Commission (2022, October). *Commission implementing regulation (EU) 2023/203 of 27 October 2022 laying down rules for the application of Regulation (EU) 2018/1139 of the European Parliament and of the Council, as regards requirements for the management of information security risks* (pp. 1–40) (Online). Available: http://data.europa.eu/eli/reg_impl/2023/203/oj

58. European Commission (2022, november). *Commission staff working document EU drone sector state of play* (pp. 1–86) (Online). Available: https://eur-lex.europa.eu/legal-content/EN/TXT/?uri=CELEX:52022SC0366
59. Brandenburger, A. M., & Stuart, H. W. (2005, January). Value-based business strategy. *Journal of Economics and Management Strategy, 5*(1), 5–24. https://doi.org/10.1111/j.1430-9134.1996.00005.x
60. Kiesewetter, L., et al. (2023, October). A holistic review of the current state of research on aircraft design concepts and consideration for advanced air mobility applications. *Progress in Aerospace Science, 142*, 100949. https://doi.org/10.1016/J.PAEROSCI.2023.100949
61. Hasan, Y. J., & Sachs, F. (2022). Performance-based preliminary design and selection of aircraft configurations for unmanned cargo operations. In *Automated low-altitude air delivery. Towards autonomous cargo transportation with drones* (pp. 107–131). Springer. https://doi.org/10.1007/978-3-030-83144-8_5
62. Hecken, T., Cumnuantip, S., & Klimmek, T. (2022). Structural Design of Heavy-Lift Unmanned Cargo Drones in low altitudes. In *Automated low-altitude air delivery. Towards autonomous cargo transportation with drones* (pp. 159–183). Springer. https://doi.org/10.1007/978-3-030-83144-8_7
63. Sliwinski, J., Gardi, A., Marino, M., & Sabatini, R. (2017). Hybrid-electric propulsion integration in unmanned aircraft. *Energy, 140*, 1407–1416. https://doi.org/10.1016/j.energy.2017.05.183
64. Rajendran, S., & Srinivas, S. (2020). Air taxi service for urban mobility: A critical review of recent developments, future challenges, and opportunities. *Transportation Research Part E Logistics and Transportation Review, 143*, 1–33. https://doi.org/10.1016/j.tre.2020.102090
65. Liu, Y., Lyu, C., Bai, F., Parishwad, O., & Li, Y. (2023, September). The role of intelligent Technology in the Development of urban air mobility systems: A technical perspective. *Fundamental Research*. https://doi.org/10.1016/J.FMRE.2023.08.006
66. Di Vito, V., et al. (2023, June). Operational concepts for urban air mobility deployment in the next decades. *Journal of Physics Conference Series, 2526*(1), 012098. https://doi.org/10.1088/1742-6596/2526/1/012098
67. ICAO (2017). *Annex 16—Environmental protection—Volume I—Aircraft noise*. Accessed: 02 December 2024 (Online). Available: https://store.icao.int/en/annex-16-environmental-protection-volume-i-aircraft-noise
68. Pundziene, A., Gutmann, T., Schlichtner, M., & Teece, D. J. (2022, August). Value impedance and dynamic capabilities: The case of MedTech incumbent-born digital healthcare platforms. *California Management Review, 64*(4), 108–134. https://doi.org/10.1177/00081256221099326
69. Zhao, R., Koo, T. T. R., Liu, W., Lodewijks, G., & Zhang, F. (2023, September). A service network design for scheduled advanced air mobility using human-driven and autonomous air metro. *Decision Analytics Journal, 8*, 100312. https://doi.org/10.1016/J.DAJOUR.2023.100312
70. Brunelli, M., Ditta, C. C., & Postorino, M. N. (2023, September). New infrastructures for urban air mobility systems: A systematic review on vertiport location and capacity. *Journal of Air Transport Management, 112*, 102460. https://doi.org/10.1016/J.JAIRTRAMAN.2023.102460
71. Wanniarachchi, S. T., & Turau, V. (2023). A study on the influence of 5G network planning on communication in urban air mobility. In *Proceedings—2023 IEEE 24th international symposium. A world wireless, mobile multimedia networks, WoWMoM 2023* (pp. 394–399). https://doi.org/10.1109/WOWMOM57956.2023.00070
72. European Union Aviation Safety Agency 2022, March. *Vertiports prototype technical specifications for the design of VFR vertiports for operation with manned VTOL-capable aircraft certified in the enhanced category (PTS-VPT-DSN)* (pp. 1–179).

73. ISO. *ISO—ISO/AWI 5491—Vertiports—Infrastructure and equipment for vertical take-off and landing (VTOL) of electrically powered cargo unmanned aircraft system (UAS).* Accessed: 20 August 2021 (Online). Available: https://www.iso.org/standard/81313.html
74. Tojal, M., Hesselink, H., Fransoy, A., Ventas, E., Gordo, V., & Xu, Y. (2022). Analysis of the definition of urban air mobility—How its attributes impact on the development of the concept. *Transportation Research Procedia, 59*, 3–13. https://doi.org/10.1016/j.trpro.2021.11.091
75. Bauranov, A., & Rakas, J. (2021). Designing airspace for urban air mobility: A review of concepts and approaches. *Progress in Aerospace Science, 125*, 100726. https://doi.org/10.1016/j.paerosci.2021.100726
76. Barrado, C., et al. (2020). U-space concept of operations: A key enabler for opening airspace to emerging low-altitude operations. *Aerospace, 7*(3), 1–18. https://doi.org/10.3390/aerospace7030024
77. Karvonen, H., Honkavaara, E., Röning, J., Kramar, V., & Sassi, J. (2023). Using a semi-autonomous drone swarm to support wildfire management—A concept of operations development study. In *Lecture notes in computer science (including subseries lecture notes in artificial intelligence and lecture notes in bioinformatics).* https://doi.org/10.1007/978-3-031-35389-5_17
78. Annex 11—Air Traffic Services|ICAO Store. Accessed: 02 December 2024 (Online). Available: https://store.icao.int/en/annex-11-air-traffic-services
79. Kramar, V., Oksman, V., Kariaho, V., & Fernbach, F. (2023). *AiRMOUR D5.2 Attractiveness of UAM EMS business models compared to alternatives, considering market potential, economic, sociological and environmental impact* (Online). Available: https://airmour.eu/uam-ems-business-models-deliverable-5-2/
80. Kramar, V., et al. (2023). *AiRMOUR D5.5 Report on operational and environmental attractiveness and sustainability of urban air mobility* (Online). Available: https://airmour.eu/operational-and-environmental-attractiveness-and-sustainability-of-urban-air-mobility-deliverable-5-5/
81. EASA (2024). *Easy access rules for unmanned aircraft systems.* Revision from July 2024. Accessed: 30 September 2022 (Online). Available: https://www.easa.europa.eu/en/downloads/110913/en
82. European Parliament and the Council (2010, October). *Regulation (EU) No 996/2010 of the European Parliament and the Council of 20 October 2010 on the investigation and prevention of accidents and incidents in civil aviation and repealing Directive 94/56/EC* (pp. 78–93) (Online). Available: https://eur-lex.europa.eu/LexUriServ/LexUriServ.do?uri=OJ:L:2010:295:0035:0050:EN:PDF
83. Hately, A., et al. (2019). *U-Space concept of operations* (Online). Available: https://www.sesarju.eu/sites/default/files/documents/u-space/CORUSConOpsvol2.pdf
84. European Commission (2019). *Commission implementing regulation (EU) 2019/317 of 11 February 2019 laying down a performance and charging scheme in the single european sky and repealing Implementing Regulations (EU) No 390/2013 and (EU) No 391/2013.*
85. European Commission (2016). *Regulation (EC) No. 562/2006 of the European Parliament and of the Council of 15 March 2006 establishing a community code on the rules governing the movement of persons across borders (Schengen Borders Code)* (pp. 1–32). 10.5040/9781782257615.0009.
86. European Commission (2020). *Directive (EU) 2020/1828 of the European Parliament and of the Council of 25 November 2020 on representative actions for the protection of the collective interests of consumers and repealing Directive 2009/22/EC* (pp. 1–27).
87. European Commission, Official website. Accessed: 30 January 2023 (Online). Available: https://commission.europa.eu/index_en
88. European Defence Agency. Accessed: 30 January 2023 (Online). Available: https://eda.europa.eu/

89. The European Organisation for Civil Aviation Equipment (EUROCAE). Accessed: 19 May 2023 (Online). Available: https://www.eurocae.net/
90. Global UTM association (GUTMA). Accessed: 19 May 2023 (Online). Available: https://gutma.org/
91. European Parliament 2008, July. *Regulation (EC) No 765/2008 of the European Parliament and of the Council of 9 July 2008 setting out the requirements for accreditation and market surveillance relating to the marketing of products and repealing Regulation (EEC) No 339/93* (pp. 30–47).
92. GSMA (2023). *MNO drone services business models.*
93. Park, J., Kim, S., & Suh, K. (2018). A comparative analysis of the environmental benefits of drone-based delivery services in urban and rural areas. *Sustainability, 10*(3). https://doi.org/10.3390/su10030888
94. Kellermann, R., Biehle, T., & Fischer, L. (2020). Drones for parcel and passenger transportation: A literature review. *Transportation Research Interdisciplinary Perspectives, 4.* https://doi.org/10.1016/j.trip.2019.100088
95. Mell, P. M., & Grance T. (2011). *The NIST definition of cloud computing.* Gaithersburg, MD. https://doi.org/10.6028/NIST.SP.800-145
96. Lenk, A., Klems, M., Nimis, J., Tai, S., & Sandholm, T. (2009). What's inside the cloud? An architectural map of the cloud landscape. In *Proceedings of the 2009 ICSE workshop on software engineering challenges of CLOUD computing, CLOUD 2009* (pp. 23–31). https://doi.org/10.1109/CLOUD.2009.5071529
97. Casadesus-Masanell, R., & Llanes, G. (2011). Mixed source. *Management Science, 57*(7), 1212–1230. (Online). Available: http://www.jstor.org/stable/25835772
98. Chesbrough, H. (2006). *Open business models: How to thrive in the new innovation landscape* (Vol. 17).
99. Helfat, C. E., Hall, T., & Raubitschek, R. S. (2018, January). Dynamic and integrative capabilities for profiting from innovation in digital platform-based ecosystems. *Dartmouth.* https://doi.org/10.2139/ssrn.3122046

The manufacturer's authorised representative in the EU is Springer Nature Customer Service Centre GmbH, Europaplatz 3, 69115 Heidelberg, Germany. If you have any concerns regarding our products, please contact ProductSafety@springernature.com

Printed and bound by CPI Group (UK) Ltd, Croydon, CR0 4YY
05/10/2026
02228021-0003